Studies in Fuzziness and Soft Computing

Volume 425

Series Editor

Janusz Kacprzyk, Systems Research Institute, Polish Academy of Sciences, Warsaw, Poland

The series "Studies in Fuzziness and Soft Computing" contains publications on various topics in the area of soft computing, which include fuzzy sets, rough sets, neural networks, evolutionary computation, probabilistic and evidential reasoning, multi-valued logic, and related fields. The publications within "Studies in Fuzziness and Soft Computing" are primarily monographs and edited volumes. They cover significant recent developments in the field, both of a foundational and applicable character. An important feature of the series is its short publication time and world-wide distribution. This permits a rapid and broad dissemination of research results.

Indexed by SCOPUS, DBLP, WTI Frankfurt eG, zbMATH, SCImago.

All books published in the series are submitted for consideration in Web of Science.

Oscar Castillo · Anupam Kumar
Editors

Recent Trends on Type-2 Fuzzy Logic Systems: Theory, Methodology and Applications

 Springer

Editors
Oscar Castillo
Research Chair of Graduate Studies
Tijuana Institute of Technology
Tijuana, Baja California, Mexico

Anupam Kumar
Department of Electronics
and Communication Engineering
National Institute of Technology
Patna, India

ISSN 1434-9922   ISSN 1860-0808 (electronic)
Studies in Fuzziness and Soft Computing
ISBN 978-3-031-26334-7   ISBN 978-3-031-26332-3 (eBook)
https://doi.org/10.1007/978-3-031-26332-3

This Springer imprint is published by the registered company Springer Nature Switzerland AG
The registered company address is: Gewerbestrasse 11, 6330 Cham, Switzerland

Preface

This book covers the introduction, theory, development and applications of type-2 fuzzy logic systems, which represent the current state of the art in various domains such as control applications, power plants, health care, image processing and mathematical applications. The book is also rich in discussing different applications in order to give the researchers a flavor of how type-2 fuzzy logic is designed for different types of problems. Type-2 fuzzy logic systems are now used extensively in engineering applications for many purposes. In simple language, this book covers the practical use of type-2 fuzzy logic and its optimization through different training methods. Furthermore, this book maintains the relationship between mathematics and practical implementations in the real world. This book chapter also contains the proper comparisons with available literature work. It shows that the presented enhanced techniques have better results. This book would serve as a handy reference guide for a variety of readers, primarily targeting research scholars, undergraduate and postgraduate researchers and practicing engineers working in Type-2 fuzzy logic systems and their applications.

Chapter Organization

Chapter 1: This chapter presents an interval type-2 fuzzy logic control-based proportional-integral-derivative (IT2-FLC-PID) method for the lower-limb exoskeleton. The proposed controller parameters are obtained using a well-known and nature-inspired optimization algorithm, cuckoo search algorithm (CSA). In order to investigate the effectiveness of the proposed technique, it is compared with the traditional type-1 FLC-PID (T1-FLC-PID) and the conventional PID control approach for exoskeleton system in health care applications.

Chapter 2: In this paper, a simplified approach to obtain mathematical models of interval type-2 (IT2) Takagi-Sugeno (TS) fuzzy PID controller using one-dimensional (1-D) input space presented. The proposed controller has been applied to nonlinear unstable processes such as continuously stirred tank reactor (CSTR) and usefulness of the proposed controller model has been checked through simulation study. It can be noted that as a model-free approach, these controllers can also be implemented for other control applications.

Chapter 3: This chapter gives a basics of fractional order calculus, its extension in fractional order PID, fractional order type 1 fuzzy and interval Type 2 controllers. The study focused on designing and developing the IT2FOFPID controller to control fractional order plants and produce better results than its conventional parts.

Chapter 4: Discussed the new two-layer interval type-2 fuzzy logic controller (TL-IT2FLC), wherein tuning the proposed control parameters is done with ABC techniques, for controlling magnetic levitation system to suspend steel ball in the air without any mechanical support. Further, proposed controller is also tested in presence of disturbance, varying reference positions, and random noise rejection to show the efficacy of the proposed and shown superior performances.

Chapter 5: Presented a novel control strategy based on Type-2 fuzzy logic sets, which can handle the uncertainties in the inputs as well as the grid disturbances in the wind energy system.

Chapter 6: In this chapter, an interval type-2 fuzzy pre-compensated PID (IT2FP-PID) controller is applied to control robotic arm and optimized for trajectory tracking problem. Here, shapes of the antecedent membership function, i.e., 60 parameters, are optimized using recent hybrid grey wolf optimizer optimization procedure. At last, the simulation results are compared with other equivalent counterparts while minimization of performance metric integral time absolute error (ITAE).

Chapter 7: Explained the Type 2 Fuzzy Logic Controller (T2FLC) design to manage the rotational speed of the oscillating water column (OWC) wave power plants. Finally, the realistic JONSWAP ocean wave model has been considered while doing the extensive simulations. The suggested T2FLC has been used to examine important OWC plant metrics such turbine power, output power, turbine flow coefficient, and rotor speed.

Chapter 8: Developed intuitionistic type-II fuzzy logic by merging type-II fuzzy logic, and intuitionistic fuzzy logic in a broader way. The proposed system can be applied in various fields including; engineering, medical and agriculture etc. So, extensive work is carried out for lung cancer patients, which consists sixteen medical entities of infected patients.

Chapter 9: This chapter aims to discuss the applicability of IoT-based enabled Type-2 Fuzzy Logic (T2FL) in the healthcare system for monitoring patients with diabetes by extracting the physiological factors from patients' bodies. A statistical experiment reveals that the model is very efficient and effective for diabetes patient monitoring, using patient risk factors.

Chapter 10: Presented a type-2 neuro-fuzzy system to deal medical problems and enhance the higher uncertainties' performance.

Chapter 11: In this chapter, the state-of-the-art literature study presented for health care sectors along with evolving of fuzzy logic. In continuation, type 2 fuzzy systems used to treat particular diseases paving the way to find further scope for further developments.

Chapter 12: This chapter presented type-2 fuzzy set-based image segmentation process to get the boundary/edge detection in blurred areas of an image. In addition, the method is verified by thermographic breast cancer image data set and the results were satisfactory.

Chapter 13: A non-iterative General Type-2 Fuzzy Logic System in quality assurance by image processing using Mamdani singleton model based on Wagner-Hagras algorithm has been described with Central Composite Design model to create a classifier. The experimental results showed that proposed model is better than the existing models.

Chapter 14: In this chapter, multiple sources of information, a large volume of data, and different parameters encourage to discuss bimatrix game problems with the type-2 fuzzy backdrop. Further, bimatrix games with payoffs of the type-2 neutrosophic fuzzy set (T2NFS) is presented. Finally, the established model is justified by offering a real-world problem to mitigate environmental pollution.

Chapter 15: In this chapter, the appropriate literature survey presented about achievement and cost expedients for discouraged arrivals Queueing System in the interval valued type-2 fuzzy environment by centroid of centroids fuzzy ordering approach. The discussed ordering approach can do this conversion effectively and provide concrete solutions.

Chapter 16: This chapter compares type-1 and type-2 fuzzy principal component analyses which are based on type1 and type2 fuzzy C-means algorithms, respectively. The two clustering methods combine the k-means clustering algorithm and type 1 and type 2 fuzzy logic, respectively.

The Main Features of the Book are as Follows

- It has covered all the latest developments and a good collection of state-of-the-art approaches of type-2 fuzzy and its applications.
- This book is very useful for the new researchers working in the field of "type-2 fuzzy logic" to quickly know the new research trends.
- The book is concisely written, lucid, comprehensive, application-based, graphical, schematics and covers wider aspects of type-2 fuzzy.

Tijuana, Mexico   Oscar Castillo
Patna, India   Anupam Kumar

Contents

About the Editors

Dr. Oscar Castillo holds the Doctor in Science degree (Doctor Habilitatus) in Computer Science from the Polish Academy of Sciences (with the Dissertation "Soft Computing and Fractal Theory for Intelligent Manufacturing"). He is a Professor of Computer Science in the Graduate Division, Tijuana Institute of Technology, Tijuana, Mexico. In addition, he is serving as Research Director of Computer Science and head of the research group on Hybrid Fuzzy Intelligent Systems. Currently, he is President of HAFSA (Hispanic American Fuzzy Systems Association) and Past President of IFSA (International Fuzzy Systems Association). Prof. Castillo is also Chair of the Mexican Chapter of the Computational Intelligence Society (IEEE). He also belongs to the Technical Committee on Fuzzy Systems of IEEE and to the Task Force on "Extensions to Type-1 Fuzzy Systems". He is also a member of NAFIPS, IFSA and IEEE. He belongs to the Mexican Research System (SNI Level 3). His research interests are in Type-2 Fuzzy Logic, Fuzzy Control, Neuro-Fuzzy and Genetic-Fuzzy hybrid approaches. He has published over 300 journal papers, 10 authored books, 50 edited books, 300 papers in conference proceedings and more than 300 chapters in edited books, in total more than 1000 publications (according to Scopus) with an h index of 80 according to Google Scholar. He has been Guest Editor of several successful Special Issues in the past, like in the following journals: Applied Soft Computing, Intelligent Systems, Information Sciences, Soft Computing, Non-Linear Studies, Fuzzy Sets and Systems, JAMRIS and Engineering Letters. He is currently Associate Editor of the Information Sciences Journal, Journal of Engineering Applications on Artificial Intelligence, International Journal of Fuzzy Systems, Journal of Complex and Intelligent Systems, Granular Computing Journal and Intelligent Systems Journal (Wiley). He was Associate Editor of the Journal of Applied Soft Computing and IEEE Transactions on Fuzzy Systems. He has been elected IFSA Fellow in 2015 and MICAI Fellow in 2016. Finally, he recently received the Recognition as a Highly Cited Researcher in 2017 and 2018 by Clarivate Analytics and Web of Science. e-mail: ocastillo@tectijuana.mx

Dr. Anupam Kumar currently serves as an Assistant Professor in the Department of Electronics and Communication Engineering, National Institute of Technology Patna, India, from 21/10/2022. Prior, he was an Assistant Professor at IIIT Kota and IIIT Bhagalpur, India, from 2019–2022 and 2018–2019, respectively. His current research interests include Fuzzy logic systems Robotics, Signal Processing, Robot Assistive device, Biomedical engineering, Health care application, Deep learning, Machine learning, etc. He received the M.Tech and Ph.D. degree from Indian Institute of Technology Roorkee, India, in 2012 and 2018 in the Electronics and Communication Engineering Department. He is a member of IEEE and has published more than 14 SCI/Scopus journals, 10 conferences, 2 book chapters and 1 patent granted. He also published 1 book titled "Tracking Performance of Maglev System using Type-2 Fuzzy Logic Control" in LAP LAMBERT Academic Publishing House. He is serving as reviewer in journals like IEEE Transactions, Elsevier, Springer, Taylor Francis journal and Wiley. He also served as a reviewer at many conferences. He is also a recipient of the best paper award at NIT Kurukshetra. He has attended several workshops and participated in many research activities. e-mail: anuanu1616@gmail.com; anupam.ec@nitp.ac.in

Application of Interval Type-2 Fuzzy Logic Control Approach to the Lower-Limb Exoskeleton

Richa Sharma and Hossein Rouhani

Abstract Fuzzy logic control methods are an integral part of control engineering and are widely used in applications such as robotics, rehabilitation, home appliances, and many more. Here, an interval type-2 fuzzy logic control-based proportional-integral-derivative (IT2-FLC-PID) control method is presented for the lower-limb exoskeleton. The intention of the use of the IT2-FLC-PID approach is to handle the system uncertainties and external disturbances encountered with the exoskeleton plant. The optimal controller parameters can be obtained using a well-known and nature-inspired optimization algorithm, the namely cuckoo search algorithm (CSA). To investigate the efficacy of the IT2-FLC-PID control scheme applied to the exoskeleton system, simulation results are compared with the traditional type-1 FLC-PID (TI-FLC-PID) and the standard PID control approach.

Keywords Interval type-2 fuzzy logic · Tracking controller · Rehabilitation · Exoskeleton · The cuckoo search algorithm

1 Introduction

It is anticipated that the elderly population older than the age of 65 years will get double from the year 1997 to 2025 [1], and it is anticipated that many of them are affected by limited mobility. Also, some conditions such as paralysis, spasticity, pain, or loss of sensation are as disabling as spinal cord injury to an individual and to society [2]. In recent times, rehabilitation robotics have gained considerable advancements for helping people with various neurological conditions. For example, the lower-limb exoskeletons are wearable rehabilitation robots that have been used to retrain these populations to restore gait toward improving the user's quality of

R. Sharma (✉) · H. Rouhani
Department of Mechanical Engineering, Donadeo Innovation Centre for Engineering, University of Alberta, Edmonton, AB T6G 1H9, Canada
e-mail: richa4@ualberta.ca

H. Rouhani
e-mail: hrouhani@ualberta.ca

© The Author(s), under exclusive license to Springer Nature Switzerland AG 2023
O. Castillo and A. Kumar (eds.), *Recent Trends on Type-2 Fuzzy Logic Systems: Theory, Methodology and Applications*, Studies in Fuzziness and Soft Computing 425,
https://doi.org/10.1007/978-3-031-26332-3_1

life and reducing the burden on healthcare systems [3, 4]. An exoskeleton generally integrates multiple technologies for sensing, actuation, and control, in the design of a wearable mechanical structure [4]. Several researchers have been working on designing and controlling different types of exoskeletons to achieve user-friendly and medically beneficial functions. Several research prototypes of wearable robots have been successfully commercialized, and their market has been rapidly evolving. However, there has been no standard or best practice or some agreed method officially available for exoskeleton devices [5].

The functioning of the rehabilitation lower-limb exoskeleton involves complicated human interaction, time-varying dynamics, and disturbances; therefore, it is challenging to design an exoskeleton along with a controller for efficient tracking of rehabilitation training tasks [3]. Several control approaches have been proposed to control an exoskeleton [6]. Fuzzy logic control (FLC) has been a widely used controller for complex nonlinear systems [3] and has the potential to be used for exoskeletons. The FLC approach has the nonlinear linguistic ability between inputs and outputs and provides a model-free and intuitive approach for complex, uncertain systems [7]. Chen et al. presented an adaptive control approach based on the disturbance-observer for a complex mechanical system with nonlinearities and validated it on a robotic exoskeleton [8]. Kiguchi et al. presented the use of neuro-fuzzy for a torque control scheme for an electromyogram-integrated robot exoskeleton [9]. Moreover, Yin et al. investigated a control approach that is adaptive and is based on a fuzzy rule interpolation method in addition to a muscle–tendon model and applied it to a robotic exoskeleton [10]. Several other authors also developed the FLC approaches for exoskeleton control [11–14].

Another emerging technique in FLC is the IT2-fuzzy logic control (IT2-FLC), which has become an excellent alternative to its traditional counterparts. The type-2 (T2) FLC approach proposed by Zadeh in 1975 can handle the uncertainties efficiently because they have membership functions that themselves are fuzzy in nature. This method seems the expansion to the standard type 1(T1) fuzzy sets. Notably, the membership grade of every element of a T2 set is itself a fuzzy set in within the limit of [0,1], whereas for a T1 fuzzy set, it is a crisp value in the range of [0,1]. It means that the T2-FLC systems have optional additional freedom to cope with and inclusion of uncertainties. Moreover, the IT2-FLC is a unique case of T2-FLC, and its use is simpler as compared to T2-FLC [15]. An IT2-FLC design provides an additional option of freedom to manage uncertainties by offering the quality of footprints of uncertainty with three-dimensional fuzzy membership functions. It has two T1-FLC membership functions, i.e., the upper, and the lower membership functions, and an additional feature footprint of uncertainty is present between both. As such, it is possible to cover a larger number of uncertainties with limited membership sets using the footprint of uncertainty feature of IT2-FLC, with the easier formulation of the rule base of the system [16]. Incorporating uncertainties in the fuzzy sets make it appropriate for applications where disturbances and noise are present in the system's performance [16]. Many recent works presented the successful use of this IT2-FLC methodology for robotic control applications [17, 18]. Therefore, the application of the IT2-FLC approach could be implemented for the control of rehabilitation exoskeleton systems. For the successful implementation of a useful control algorithm, the most favorable values of controller parameters can be obtained by various meta-heuristic and advanced optimization practices. In the presented chapter, an effective method named as cuckoo search algorithm (CSA) is utilized as an optimization tool to get the ideal values of controller gains for the discussed IT2-FLC scheme. In recent years, the method CSA has been gaining extreme popularity for discovering the optimum parameters for different controller techniques for robots [19, 20] and lower-limb prostheses [21]; therefore, it could also be a potential technique for control of lower-limb exoskeletons.

The key contribution of this chapter is to illustrate the applicability of the IT2-FLC-PID controller to the lower-limb exoskeleton as a complex and uncertain plant. The ultimate goal of use of this control approach is to manage the external disturbances and uncertainties present during the execution of any motor task. The CSA is used as a tool to obtain optimized controller gains for the proposed method. To investigate the successful application of discussed IT2-FLC-PID control approach, its performance is assessed with its counterpart T1-FLC-PID and traditional proportional-integral-derivative (PID) controller.

2 Mathematical Representation of a Lower-Limb Exoskeleton

The mathematical model representation of the used lower-limb exoskeleton plant considered for this work [14, 22], is presented here. This mathematical model provides a relationship between the link positions (θ_{LLE_i}) and the external force or torque (τ_{LLE_i}) of a lower-limb exoskeleton displayed in Fig. 1 [14]. Also, the mathematical relation is presented by:

$$
\begin{aligned}
(M_{LLE}(\theta_{LLE_i}) &+ \bar{M}_{LLE}(\theta_{LLE_i}))\ddot{\theta}_{LLE_i} + (CF_{LLE}(\theta_{LLE_i}, \dot{\theta}_{LLE_i}) \\
&+ \overline{CF}_{LLE}(\theta_{LLE_i}, \dot{\theta}_{LLE_i}))\dot{\theta}_{LLE_i} + G_{LLE}(\theta_{LLE_i}) + \bar{G}_{LLE}(\theta_{LLE_i}) \\
&= \tau_{LLE_i} + F_{LLE_i}(t) + \Delta(t_e, \theta_{LLE_i}, \dot{\theta}_{LLE_i}, \ddot{\theta}_{LLE_i})
\end{aligned}
\tag{1}
$$

The inertia matrix is given by

$$
M_{LLE}(\theta_{LLE_i}) = \begin{bmatrix} M_{L_{11}} & M_{L_{12}} \\ M_{L_{21}} & M_{L_{22}} \end{bmatrix}
\tag{2}
$$

$$
M_{L_{11}} = m_{lle_1}d_{lle_1}^2 + m_{lle_2}(l_{lle_1}^2 + d_{lle_2}^2 + 2l_{lle_1}d_{lle_2}\cos(\theta_{LLE_2})) + I_{lle_1} + I_{lle_2}
\tag{3}
$$

$$
M_{L_{12}} = m_{lle_2}(d_{lle_2}^2 + l_{lle_1}d_{lle_2}\cos(\theta_{LLE_2})) + I_{lle_2}
\tag{4}
$$

Fig. 1 Basic design of
lower-limb exoskeleton

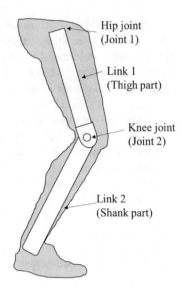

Hip joint
(Joint 1)

Link 1
(Thigh part)

Knee joint
(Joint 2)

Link 2
(Shank part)

$$M_{L_{21}} = M_{L_{12}} \tag{5}$$

$$M_{L_{22}} = m_{lle_2} d_{lle_2}^2 + I_{lle_2} \tag{6}$$

In this, m_{lle_1} (2.6 kg) and m_{lle_2} (3.2 kg) state the body mass of Link 1 and Link 2,; l_{lle_1} (25 cm) and l_{lle_2} (30 cm) correspond to the total length of the Link 1 and the Link 2, respectively; τ_{LLE_1} as well as τ_{LLE_2} signify the external torques/forces on Link 1 and Link 2, respectively; $I_{lle_1} = \frac{1}{3} m_{lle_1} l_{lle_1}^2$ and $I_{lle_2} = \frac{1}{12} m_{lle_2} l_{lle_2}^2$ evaluate the moment of inertia for the Link 1 as well as the Link 2; d_{lle_1} (12 cm) and d_{lle_2} (15 cm) correspond to the distance from the joint end to the centre of mass of the Link 1 as well as the Link 2; and g represents the acceleration due to gravitation [14].

Further, the Coriolis and centrifugal force matrix is expressed as

$$CF_{LLE}(\theta_{LLE_i}, \dot{\theta}_{LLE_i}) = \begin{bmatrix} CF_{lle_{11}} & CF_{lle_{12}} \\ CF_{lle_{21}} & CF_{lle_{22}} \end{bmatrix} \tag{7}$$

$$CF_{lle_{11}} = -m_{lle_2} l_{lle_1} d_{lle_2} \dot{\theta}_{LLE_2}^2 \sin(\theta_{LLE_2}) \tag{8}$$

$$CF_{lle_{12}} = -m_{lle_2} l_{lle_1} d_{lle_2} (\dot{\theta}_{LLE_1} + \dot{\theta}_{LLE_2}) \sin(\theta_{LLE_2}) \tag{9}$$

$$CF_{lle_{21}} = m_{lle_2} l_{lle_1} d_{lle_2} \dot{\theta}_{LLE_1} \sin(\theta_{LLE_2}) \tag{10}$$

$$CF_{lle_{22}} = 0 \tag{11}$$

The gravity vector is presented below

$$G_{LLE}(\theta_{LLE_i}) = \begin{bmatrix} G_{lle_{11}} \\ G_{lle_{21}} \end{bmatrix} \tag{12}$$

$$G_{lle_{11}} = \left(m_{lle_1}d_{lle_2} + m_{lle_2}l_{lle_1}\right)g\cos(\theta_{LLE_1}) \\ + m_{lle_2}d_{lle_2}g\cos(\theta_{LLE_1} + \theta_{LLE_2}) \tag{13}$$

$$G_{lle_{21}} = m_{lle_2}d_{lle_2}g\cos(\theta_{LLE_1} + \theta_{LLE_2}) \tag{14}$$

$$\zeta_{lle}(\theta_{LLE_i}, \dot{\theta}_{LLE_i}, \ddot{\theta}_{LLE_i}) \triangleq \bar{M}_{LLE}(\theta_{LLE_i})\ddot{\theta}_{LLE_i} \\ + \overline{CF}_{LLE}(\theta_{LLE_i}, \dot{\theta}_{LLE_i})\dot{\theta}_{LLE_i} \\ + \bar{G}_{LLE}(\theta_{LLE_i}) = 0.2(M_{LLE}(\theta_{LLE_i})\ddot{\theta}_{LLE_i} \\ + CF_{LLE}(\theta_{LLE_i}, \dot{\theta}_{LLE_i})\dot{\theta}_{LLE_i} + G_{LLE}(\theta_{LLE_i})) \tag{15}$$

The interaction between used lower-limb exoskeleton and the wearer are characterized as

$$F_{LLE_1}(t) = 2\cos(2\pi t) \tag{16}$$

$$F_{LLE_2}(t) = 2\sin(2\pi t) \tag{17}$$

where $e_{lle_i}(t)$ symbolizes the tracking difference between the reference and the measured trajectory with $i = 1, 2$ depicted the Joint 1 and the Joint 2.

Moreover, the frictional forces, as well as present disturbances in the plant, are denoted as [14, 22].

$$\Delta(t_e) = 1 + 3\|\dot{e}_{lle_i}(t)\| + 2\|e_{lle_i}(t)\| \tag{18}$$

3 Controller Strategy

In this section of the chapter, a control strategy IT2-FLC-PID for a lower-limb exoskeleton, is presented. Also, the optimization technique employed for getting the controller gains is briefly discussed.

3.1 An Interval Type-2 Fuzzy Logic PID Controller (IT2-FLC-PID)

The fundamental design and implementation of the IT2-FLC-PID approach as demonstrated in Fig. 2 is discussed in this section.

A standard PID controller is stated as:

$$U_{PID_i}(t) = K_{pl_i} e_{lle_i}(t) + K_{il_i} \int e_{lle_i}(t)dt + K_{dl_i} \frac{de_{lle_i}(t)}{dt} \qquad (19)$$

where K_{pl_i}, K_{dl_i} K_{il_i} denote the proportional gain, the derivative, and integral gain, of a standard PID control method.

An intelligent control method FLC offers a nonlinear mapping among the input variables and output variable quantities to the controller. Here, the FLC-PID controller involves the aggregation of the FLC-PI as well as the FLC-PD controller methods. This input vs. output variables mapping for designing the control approach is expressed by the following equation [14], and more details are available in [7, 14]:

$$\frac{T}{(1 - z^{-1})} \alpha_{lle_i} \Delta u_{PI_{LLE}}(k) + \beta_{lle_i} u_{PD_{LLE}}(k) = k_{lle_i} e_{lle_i}(k) + q_{lle_i} \frac{(1 - z^{-1})}{T} e_{lle_i}(k) \qquad (20)$$

where α_{lle_i}, β_{lle_i}, k_{lle_i}, and q_{lle_i} ($i = 1, 2$) are the scaling parameters of the stated IT2-FLC-PID control technique for each joint of used exoskeleton.

The IT2-FLC-PID controller structure as shown in Fig. 3 consists of same components as the T1-FLC-PID structure that are the fuzzification process, core inference engine, created rule base, and finally, the defuzzification process. Also, the only additional component in IT2-FLC is the type-reducer [23]. The type-reducer converts the T2 fuzzy sets to T1 fuzzy sets, and then these T1-fuzzy sets are plugged into the defuzzification block to convert the fuzzy values into crisp values [15, 16].

The significant feature in the IT2-FLC-PID controller is its membership functions. A single membership function of IT2-FLC consists of two T1 membership functions,

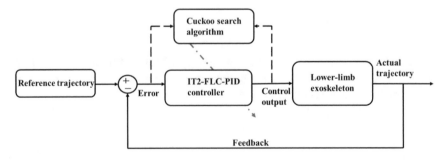

Fig. 2 Fundamental scheme of used IT2-FLC-PID control approach to a lower-limb exoskeleton

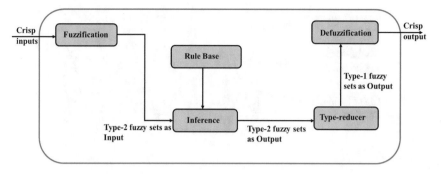

Fig. 3 Design structure of IT2-FLC-PID controller

i.e., the upper membership function, and the lower membership functions, and in between these two, the additional footprint of uncertainty lies as indicated in Fig. 4 (a). Assume that an antecedent membership function can be represented by $\tilde{X}_{it2}$ and the following consequent membership function is represented by $\tilde{Y}_{it2}$ therefore, the upper membership functions are represented as $\left(\bar{\mu}_{\tilde{X}'_{it2}}, \bar{\mu}_{\tilde{Y}'_{it2}} \right)$, and lower membership functions are represented as $\left(\underline{\mu}_{\tilde{X}'_{it2}}, \underline{\mu}_{\tilde{Y}'_{it2}} \right)$, and the combined firing rule is given as follows [16]:

$$\widetilde{ff}_{it2} = [\underline{ff}^{it2}, \overline{ff}^{it2}] \tag{21}$$

where $\underline{ff}^{it2} = \left(\underline{\mu}_{\tilde{X}'_{it2}} * \underline{\mu}_{\tilde{Y}'_{it2}} \right); \overline{ff}^{it2} = \left(\bar{\mu}_{\tilde{X}'_{it2}} * \bar{\mu}_{\tilde{Y}'_{it2}} \right)$

Now, for the type-reducing of output, the centre of sets method is utilized and then defuzzifier is applied as [16]:

$$u_{it2-flc} = \left[\frac{u_{ll_{flc}} + u_{rr_{flc}}}{2} \right] \tag{22}$$

where $u_{ll_{flc}} = \dfrac{\sum_{n=1}^{ll} \overline{ff}^{it2} C^n + \sum_{n=ll+1}^{N} \underline{ff}^{it2} C^n}{\sum_{n=1}^{ll} \overline{ff}^{it2} + \sum_{n=ll+1}^{N} \underline{ff}^{it2}};$

$$u_{rr_{flc}} = \frac{\sum_{n=1}^{rr} \underline{ff}^{it2} C^n + \sum_{n=rr+1}^{N} \overline{ff}^{it2} C^n}{\sum_{n=1}^{rr} \underline{ff}^{it2} + \sum_{n=rr+1}^{N} \overline{ff}^{it2}}$$

In the above expressions, the switching points (ll, rr) are attained with well-known Karnik–Mendel type-reduction method [16, 24].

In this chapter, the used key input variables are the tracking error/difference and its derivative, and the output variable quantity is control torque output, i.e., torque exerted by the actuators on the joints. Here, membership functions used for both input

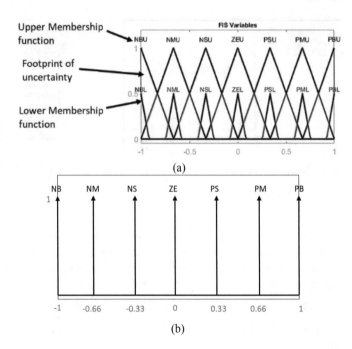

Fig. 4 Used Membership functions **a** input variables (tracking difference/error and derivative of tracking difference) **b** the output variables (control torque output) for IT2-FLC-PID controller

variables are seven conventional triangular membership functions as indicated in Fig. 4a whereas the stated output variables are seven singleton membership functions displayed in Fig. 4b that are: Negative Big (NB), Positive Big (PB). Negative Medium (NM), Positive Medium (PM), Negative Small (NS), Positive Medium (PM) and Zero (ZE). Using the mentioned input and output features, the 7 × 7 rule base is prepared provided in Table 1 [16] which is used to generate the appropriate control action.

3.2 Cuckoo Search Algorithm (CSA)

For this work, the CSA is used as a tool to obtain optimized controller gains or parameters as needed for the implementation of an effective control method. The mentioned CSA was founded by Yang and Deb in the year 2009. It runs on the parasitic breeding actions of the famous cuckoo birds [25, 26]. A detailed description of this algorithm can be found in [27, 28]. Additionally, the major rules for implementation of this method are stated as follows [26–28]:

Each cuckoo bird can lay only an egg and place that in an aimlessly searched nest.

(a) The calculated best nest should contain the optimum solution as well as it must keep this to all future group (iterations).

Table 1 Rule base formulated between the stated input as well as output variables for the mentioned IT2-FLC-PID and T1-FLC-PID control techniques

Tracking difference/error

		NB	NM	NS	ZE	PS	PM	PB
Derivative of the tracking difference/error	NB	NB	NB	NB	NB	NM	NS	ZE
	NM	NB	NB	NB	NM	NS	ZE	PS
	NS	NB	NB	NM	NS	ZE	PS	PM
	ZE	NB	NM	NS	ZE	PS	PM	PB
	PS	NM	NS	ZE	PS	PM	PB	PB
	PM	NS	ZE	PS	PM	PB	PB	PB
	PB	ZE	PS	PM	PB	PB	PB	PB

PB: Positive Big NB: Negative Big, PM: Positive Medium, NM: Negative Medium,
PS: Positive Small, NS: Negative Small, and ZE: Zero

(b) The stated host nests are always fixed in number, and a host (presenter) bird would detect an invader with a probability of [0,1].

For the implementation of CSA, it is essential to select an appropriate performance index. For this chapter, cost function used for this algorithm is the sum of integral of absolute difference/error (IAE) of both joints and integral absolute change in controller output torque [7].

The general steps involved in CSA is presented below [26–28]:

I.   At random generate a population of G_{csa} host cuckoo birds is generated, set up the number of iterations (generations) $= 100$.
II. Create a cuckoo search with the Leúy flight. Calculate its fitness value f_c.
III. Randomly decided a nest u from the host residents. Calculate the fitness value f_u.
IV. Finally, decide if $f_c < f_u$; if true, substitute u with a new candidate; if false, keep this as the new solution.
V.   Abandon a fraction of birds to avoid raiders with the likelihood of 0.25. Create a brand-new nest.
VI. Preserve the current best solution. Check the generation and increment the counter.
VII. Repeat steps II–VI till the final generation reaches.

4   Results and Discussions

In this segment, simulation results for the performance assessment of the application of IT2-FLC-PID controller to a mentioned lower-limb exoskeleton plant for a tracking job is presented. The comparison of the proposed controller is investigated for its counterpart T1-FLC-PID method as well as the standard PID controller. The MATLAB/Simulink platform is utilized for implementing this work. Also, the open-source IT2-FLC toolbox [29] is used to implement the IT2-FLC-PID control approach. The desired trajectories for the Joint 1 and the Joint 2 of stated lower-limb exoskeleton are provided in Eqs. (23) and (24) below:

$$\theta_{ref_1}(t) = 0.8 + 0.2\sin(2\pi t)rad \tag{23}$$

$$\theta_{ref_2}(t) = 1.2 - 0.2\cos(2\pi t)rad \tag{24}$$

For this work, the initial tracking errors for Joint 1 and Joint 2 are set as $e_{lle_1}(0) = 0.1$ and $e_{lle_2}(0) = 0.15$ radians [14, 22]. As discussed in the above section, the controller parameters for all control schemes are optimized using the CSA. There are four parameters for each joint that need to be optimized. The optimized parameters and resulting IAE values are detailed in Table 2. Additionally, the subsequent IAE values obtained for the IT2-FLC-PID are 0.007684 and 0.0226 for the Joint 1 and Joint 2, respectively. These values are smaller than the IAE values of the other mentioned controller methods, the conventional T1-FLC-PID, and standard PID controllers. The IAE values for T1-FLC-PID controller are 0.01242 and 0.06141 for Joint 1 and Joint 2, respectively, whereas the IAE values for PID controllers are 0.06253 and 0.1091 for the Joint 1 and the Joint 2. Also, the tracking performance of proposed control schemes for the Joint 1 and the Joint 2 are displayed in Fig. 5a, b. Furthermore, the controller outputs of all schemes for Joint 1 and Joint 2 of the lower-limb exoskeleton are represented in Fig. 5c, d. Moreover, the tracking error performance of the Joint 1 and the Joint 2 for all control schemes are demonstrated in the Fig. 5e, f. From Table 2 and Fig. 5e, f, it can be inferred that the tracking errors for the IT2-FLC-PID control scheme are smaller as assessed to the stated two controllers, and it signifies the superiority of IT2-FLC-PID over its counterpart T1-FLC-PID method and standard PID controllers in terms of tracking performance. Therefore, this chapter provides the application of IT2-FLC-PID control to the complex lower-limb exoskeleton plant for tracking performance. In this work, the external disturbances and uncertainties have already been present in the system under consideration which indicates that the proposed IT2-FLC-PID outperforms other controller, i.e., T1-FLC-PID and the traditional PID controller for trajectory tracking task for uncertain and complex lower-limb exoskeleton plant. However, there are a few limitations of this study. This study presents the simulation results for the stated control scheme employed to a lower-limb exoskeleton, which does not include uncertainties and disturbances present in the real world that may have an adverse effect on the performance of the plant. In future works, the real-time implementation of the suggested IT2-FLC-PID controller will be investigated in the existence of external disturbances and noises.

Table 2 Parameters obtained using CSA and subsequent IAE for the stated IT2-FLC-PID, T1-FLC-PID, and standard PID control techniques

IT2-FLC-PID				T1-FLC-PID				PID			
Joint 1		Joint 2		Joint 1		Joint 2		Joint 1		Joint 2	
Parameters	Values	Parameters	Values	Parameters	Values	Parameters	Values	Parameters	Values	Parameters	Values
k_{lle_1}	189.6513	k_{lle_2}	32.2774	k_{lle_1}	70.0002	k_{lle_2}	105.2009	K_{pl_1}	309.3560	K_{pl_2}	198.9001
q_{lle_1}	2.4000	q_{lle_2}	0.0010	q_{lle_1}	0.0956	q_{lle_2}	1.2760	K_{il_1}	46.7894	K_{il_2}	16.0978
β_{lle_1}	14.0007	β_{lle_2}	25.2010	β_{lle_1}	30.0235	β_{lle_2}	2.7845	K_{dl_1}	25.6757	K_{dl_2}	56.9806
α_{lle_1}	0.8000	α_{lle_2}	0.0100	α_{lle_1}	2.8967	α_{lle_2}	0.0020	–		–	–
IAE	0.007684	IAE	0.0226	IAE	0.01242	IAE	0.06141	IAE	0.06253	IAE	0.1091

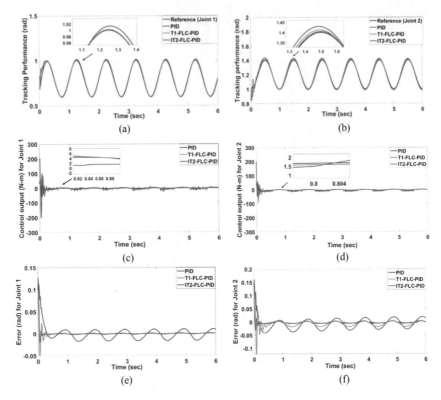

Fig. 5 **a** The sinusoidal trajectory tracking outcome for the Joint 1; **b** the trajectory tracking outcome for the Joint 2; **c** control torque output for the Joint 1; **d** control torque output for the Joint 2; **e** tracking error (difference) for the Joint 1; **f** tracking error (difference) for the Joint 2 of a stated lower-limb exoskeleton for standard PID, conventional TI-FLC-PID, and stated IT2-FLC-PID control techniques

5   Conclusions

In this chapter, a IT2-FLC-PID control technique is employed for the control of a lower-limb exoskeleton plant. The optimized controller gains for this controller scheme have been obtained using a CSA optimization technique. The simulation outcomes are then compared to other control techniques, namely conventional T1-FLC-PID and standard PID controller. From the performance outcome, it is concluded that the IT2-FLC-PID control approach outperformed the other two controllers in terms of tracking errors and was able to deal with the disturbances and plant uncertainties. Overall, this chapter provides an idea of simulation work on the applicability and suitability of discussed IT2-FLC-PID controller method for a lower-limb exoskeleton plant for the tracking job. In future, the real-time execution of the stated control scheme should be the investigated. Moreover, modern control techniques such as adaptive control, fractional order control method and intelligent control like neural network based-control could also be explored for the tracking job of a lower-limb exoskeleton.

References

1. Cooper, R.A., Dicianno, B.E., Brewer, B., LoPresti, E., Ding, D., Simpson, R., Grindle, G., Wang, H.: A perspective on intelligent devices and environments in medical rehabilitation. Med. Eng. Phys. **30**, 1387–1398 (2008)
2. Zeilig, G., Weingarden, H., Zwecker, M., Dudkiewicz, I., Bloch, A., Esquenazi, A.: Safety and tolerance of the Rewalk™ exoskeleton suit for ambulation by people with complete spinal cord injury: a pilot study. J. Spinal Cord Med. **35**(2), 96–101 (2012)
3. Aliman, N., Ramli, R., Haris, S.M., Amiri, M.S., Van, M.: A robust adaptive fuzzy-proportional-derivative controller for a rehabilitation lower limb exoskeleton. Eng. Sci. Technol. Int. J. **35**, 101097 (2022)
4. Shi, D., Zhang, W., Zhang, W., Ding, X.: A review on lower limb rehabilitation exoskeleton robots. Chinese J. Mech. **32**(74), 1–11 (2019)
5. Pinto-Fernandez, D., Torricelli, D., Sanchez-Villamanan, M.D.C., Aller, F., Mombaur, K., Conti, R., Vitiello, N., Moreno, J.C., Pons, J.L.: Performance evaluation of lower limb exoskeleton: a systematic review. IEEE Trans. Neural Syst. Rehabil. Eng. **28**(7), 1573–1583 (2020)
6. Baud, R., Manzoori, A., Ljspeert, A., Bouri, M.: Review of control strategies for lower-limb exoskeletons to assist gait. J. Neuro Eng. Rehabil **18**, 119 (2021)
7. Sharma, R., Rana, K.P.S., Kumar, V.: Performance analysis of fractional order fuzzy PID controllers applied to a robotic manipulator. Expert Syst. Appl. **41**(9), 4274–4289 (2014)
8. Chen, Z., Li, Z., Philip Chen, C.L.: Disturbance observer-based fuzzy control of uncertain MIMO mechanical systems with input nonlinearities and its application to robotic exoskeleton. IEEE Trans. Cybern. **47**(4), 984–994 (2017)
9. Kiguchi, K., Iwami, K., Yasuda, M., Watanabe, K.: An exoskeleton robot for human shoulder joint motion assist. IEEE/ASME Trans. Mechatron. **8**(1), 125–135 (2003)
10. Yin, K., Xiang, K., Pang, M., Chen, J., Anderson, P., Yang, L.: Personalised control of robotic ankle exoskeleton through experience-based adaptive fuzzy inference. IEEE Access **7**, 72221–72233 (2019)
11. Niu, J., Song, Q., Wang, X.: Fuzzy PID control for passive lower extremity exoskeleton in swing phase. In: 2013 IEEE 4th International Conference on Electronics, Information and Emergency Communication, 15–17 Nov 2013, Beijing, China
12. Kong, K., Jeon, D.: Fuzzy control of a new tendon-driven exo-skeletal power assistive device. In: Proceeding, 2005 IEEE/ASME International Conference Advance Intelligent Mechatronics IEEE 2005, pp. 146–151 (2005)
13. Rezage, G.Al., Tokhi, M.O.: Fuzzy PID control of lower limb exoskeleton for elderly mobility. Proceedings AQTR: IEEE International Conf. on Automation, Quality and Testing, Robotics, Cluj-Napoca, Romania, 19–21 May 2016, pp. 1–6
14. Sharma, R., Gaur, P., Bhatt, S., Joshi, D.: Optimal fuzzy logic-based control strategy for lower-limb rehabilitation exoskeleton. Appl. Soft Comput. **105**, 107226 (2021)
15. El-Bardini, M., El-Nagar, A.M.: Interval type-2 fuzzy PID controller for uncertain nonlinear inverted pendulum system. ISA Trans. **53**, 732–743 (2014)
16. Sharma, R., Deepak, K.K., Gaur, P., Joshi, D.: An optimal interval type-2 fuzzy logic control based closed-loop drug administration to regulate the mean arterial blood pressure. Comput. Methods Programs Biomed. **185**, 105167 (2020)

17. Gaidhane, P.J., Nigam, M.J., Kumar, A., Pradhan, P.M.: Design of interval type-2 fuzzy logic precompensated PID control applied to two-DOF robotic manipulator with variable payload. ISA Trans. **89**, 169–185 (2019)
18. Kumar, A., Kumar, V.: Design of interval type-2 fractional order fuzzy logic controller for redundant robot with artificial bee colony. Arabian J. Sci. Eng. **44**, 1883–1902 (2019)
19. Sharma, R., Gaur, P., Mittal, A.P.: Performance analysis of two-degree of freedom fractional order PID controllers for robotic manipulator with payload. ISA Trans. **58**, 279–291 (2015)
20. Sharma, R., Gaur, P., Mittal, A.P.: Design of two-layered fractional order fuzzy logic controllers applied to robotic manipulator with variable payload. Appl. Soft Comput. **47**, 565–576 (2016)
21. Sharma, R., Gaur, P., Bhatt, S., Joshi, D.: Performance assessment of fuzzy logic control approach for MR-damper based-transfemoral prosthetic leg. IEEE Trans. Artif. Intell. **3**(1), 53–66 (2021)
22. Yang, Y., Huang, D., Dong, X.: Enhanced neural network control of lower limb rehabilitation exoskeleton by add-on repetitive learning. Neurocomputing **123**, 256–264 (2019)
23. Castillo, O., Melin, P.: A review on the design and optimization of interval type-2 fuzzy controllers. Appl. Soft Comput. **12**, 1267–1278 (2012)
24. Liang, Q., Mendel, J.M.: Interval type-2 fuzzy logic systems: theory and design. IEEE Trans. Fuzzy Syst. **8**, 535–550 (2000)
25. Yang, X.S., Deb, S.: Cuckoo search via Leúy flight. In: Proceedings of the World Congress on Nature & Biologically Inspired Computing, India, pp. 210–214 (2009)
26. Sharma, R., Bhasin, S., Gaur, P., Joshi, D.: A switching-based collaborative fractional order fuzzy logic controllers for robotic manipulators. Appl. Math. Model. **73**, 228–246 (2019)
27. Sharma, R., Gaur, P., Mittal, A.P.: Optimum design of fractional-order hybrid controller for a robotic manipulator. Arabian J. Sci. Eng. **42**(2), 739–750 (2017)
28. Yang, X.S., Deb, S.: Engineering optimization by cuckoo search. Inte. J. Math. Model. Numer. Optim. **1**(4), 330–343 (2010)
29. Taskin, A., Kumbasar, T.: An open source Matlab/Simulink toolbox for interval type-2 fuzzy logic systems. In: Proceedings of the IEEE Symposium Series on Computational Intelligence, pp. 561–568 (2015)

A Simplified Model of an Interval Type-2 Takagi-Sugeno Fuzzy PID Controller using One-Dimensional Input Space

Ritu Raj, Anupam Kumar, and Prashant Gaidhane

Abstract Deriving analytical structures or mathematical models are quite common to reveal the black-box nature of fuzzy controllers. However, this approach was applied on a two-/three- dimensional input space resulting in complex models of fuzzy controllers. In this paper, we propose a simplified approach to obtain mathematical models of interval type-2 (IT2) Takagi-Sugeno (TS) fuzzy PID controller using one-dimensional (1-D) input space. The fuzzy PID control action is designed by individually modelling and combining fuzzy proportional (F-P), fuzzy integral (F-I), and fuzzy derivative (F-D) control actions on 1-D input space. In this approach, the need of triangular norms and co-norms is eliminated, resulting in a simplified controller model. To show the usefulness of the proposed controller model, we present simulation studies on nonlinear unstable processes such as continuously stirred tank reactor (CSTR). However, as a model free approach, these controllers can also be implemented for other control applications.

Keywords Fuzzy control · Analytical structure · modeling · Nonlinear control · CSTR

1 Introduction

Fuzzy logic systems (FLS) have become quite popular in the last several decades due to its capability to handle well-defined and ill-defined processes [1]. One of the first applications of FLS was in the domain of control systems [2]. However, FLS

R. Raj (✉)
Indian Institute of Information Technology Kota, MNIT Campus Jaipur, Jaipur, India
e-mail: rituraj.ece@iiitkota.ac.in

A. Kumar
National Institute of Technology Patna, Patna, Bihar, India
e-mail: anupam.ec@nitp.ac.in; anuanu1616@gmail.com

P. Gaidhane
Government College of Engineering, Jalgaon, India

© The Author(s), under exclusive license to Springer Nature Switzerland AG 2023
O. Castillo and A. Kumar (eds.), *Recent Trends on Type-2 Fuzzy Logic Systems: Theory, Methodology and Applications*, Studies in Fuzziness and Soft Computing 425,
https://doi.org/10.1007/978-3-031-26332-3_2

based controllers were treated as black-box whose closed-form solutions were not available. In early 1990s fuzzy controllers were developed to tackle this problem. Ying [3] presented the simplest fuzzy PI controllers revealing their mathematical structures. Later this approach was extended to fuzzy PD and lately to fuzzy PID controllers.

Mizumoto [4] realized fuzzy PID controllers using "product-sum-gravity method" and "simplified fuzzy reasoning method". It was also shown that fuzzy PID controllers cannot be constructed using min-max-gravity method. Misir et al. [5] presented a design method and stability analysis of a fuzzy PI plus derivative controller. Later, Mann et al. [6] presented different fuzzy PID controller structure by manipulating the rule base in different forms. Golob [7] further highlighted some simplified structures of fuzzy PID controllers. These controller structures used "three one-input one-output inferences with three separate rule bases". Recently, type-1 fuzzy PID controllers of Takagi-Sugeno (TS) type [8] and Mamdani type [9] were developed using three inputs resulting in a three-dimensional input space. Raj et al. [10] simplified the structure of a interval type-2 (IT2) fuzzy PID controller by utilizing a two-dimensional input space. However, this structure of fuzzy PID controller used three inputs. Further, Sain and Mohan [11] developed type-1 fuzzy PID controllers using one-dimensional (1-D) input space which reduced the computational complexity of the controller. In this chapter we extend the work in [11] to an IT2 fuzzy PID controller using 1-D input space.

Apart from integer order fuzzy controllers, several works on fractional order fuzzy controllers have been reported in the literature. A fractional order fuzzy PID controller was proposed [12] where the controller parameters were suitably designed using genetic algorithm based optimization. For uncertain systems such as two-link manipulators, Kumar et al. [13] proposed a fractional order fuzzy proportional (F-P) plus fuzzy integral (F-I) plus fuzzy derivative (F-D) controller. The controller parameters were again tuned using genetic algorithm based optimization. Recently, Kumar and Raj [14] presented a design method for a fractional order two-layer fuzzy controller for controlling the mean arterial blood pressure. It is important to note that the fuzzy controllers do not have an unique model. It depends upon the proper selection of membership functions, rule base, inference mechanism, type-reduction and defuzzification.

The objective of this work is to present a simplified modelling approach for IT2 TS fuzzy PID controller where each control action i.e., F-P, F-I, and F-D is realized individually on a 1-D input space. The control actions are then merged to provide the PID control law. A 1-D input space eliminates the need of triangular norms and co-norms for deriving the controller model, hence, reducing the complexity. Most often it is found in the literature [8–10, 15–21] that the controller model utilizes a two-/three- dimensional input space (or state-space) resulting in complex controller models [22]. Some works [23, 24] have utilised the parallel combination of F-P/F-I/F-D control action to yield a fuzzy PID controller. However, the outputs of F-P/F-I/F-D controllers are nonlinear functions of error and change of error in [23], and the output of F-P controller is a nonlinear function of error and change of error, and the outputs of F-I/F-D controllers are nonlinear functions of error and rate of change of error in [24]. Whereas the outputs of F-P, F-I, and F-D controllers are nonlinear functions of change of error, error, and double change of error, respectively, in this work. Moreover, a total of nine [23] and four [24] rules have been employed for F-P, F-I, and F-D controllers, whereas only two rules have been considered in this work. A similar approach has been reported for a type-1 Mamdani fuzzy PID controller using 1-D input space [11]. However, it will be proven in this work that type-1 fuzzy PID controllers are a special case of IT2 fuzzy PID controllers.

This chapter consists of five sections: Sect. 1 presents the background and literature survey relevant to this work. Section 2 discusses about conventional and fuzzy PID controller theory. In Sect. 3 the mathematical model of IT2 TS fuzzy PID controller in 1-D input space is derived. Section 4 presents the simulation study to validate the controller structure. Finally, Sect. 5 concludes this chapter.

2 PID Controller: Conventional and Fuzzy

A continuous-time system can be controlled by a discrete-time controller as shown in Fig. 1. The zero-order hold (ZOH) converts the discrete-time control signal $u(k)$ into continuous-time control signal $u(t)$. $r(t)$, $e(t)$, $e(k)$, $d_i(t)$, $d_o(t)$, $y(t)$, and T denote the reference signal, continuous-time error signal, error signal at kth sampling instant, disturbance signal applied at input and output of the plant, plant response, and sampling time, respectively. The control signal $u^{PID}(t)$ of a continuous-time linear PID controller in terms of error signal $e(t)$ is given as follows:

$$u^{PID}(t) = K_P^c e(t) + K_I^c \int_0^t e(\tau)d\tau + K_D^c \frac{de(t)}{dt} \tag{1}$$

where K_P^c, K_I^c, and K_D^c denote proportional, integral, and derivative gains of continuous-time linear PID controller. Differentiating Eq. (1) w.r.t t yields

$$\frac{du^{PID}(t)}{dt} = K_P^c \frac{de(t)}{dt} + K_I^c e(t) + K_D^c \frac{d^2 e(t)}{dt^2} \tag{2}$$

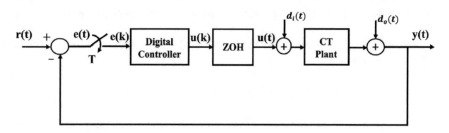

Fig. 1 A typical closed-loop computer control system [11]

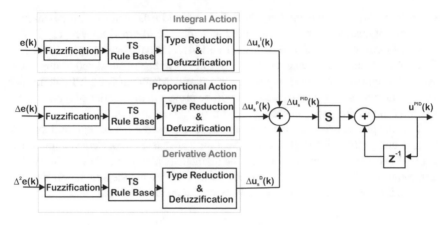

Fig. 2 Proposed block diagram of type-2 TS fuzzy PID controller using 1-D input space

At k^{th} sampling instant, Eq. (2) can be expressed as

$$\Delta u^{PID}(k) = K_P^d \Delta e(k) + K_I^d e(k) + K_D^d \Delta^2 e(k) \tag{3}$$

where $e(k)$, $\Delta e(k)$, and $\Delta^2 e(k)$ denote error signal, change of error signal, and double change of error signal, respectively. K_P^d, K_I^d, and K_D^d are the proportional, integral, and derivative gains of discrete-time linear PID controller. $\Delta u^{PID}(k)$ is the incremental PID control action. The final PID control action of a conventional discrete-time PID controller is given as

$$u^{PID}(k) = u^{PID}(k-1) + \Delta u^{PID}(k) \tag{4}$$

One can see from Eq. (3), that the incremental PID control action can be generated by individually summing the proportional, integral, and derivative action with respect to the inputs $\Delta e(k)$, $e(k)$, and $\Delta^2 e(k)$. Based on this approach, the block diagram of a type-2 TS fuzzy PID controller is presented in Fig. 2. The fuzzy control actions, i.e. fuzzy proportional ($\Delta u_s^P(k)$), fuzzy integral ($\Delta u_s^I(k)$), and fuzzy derivative ($\Delta u_s^D(k)$), are computed individually and summed together to arrive at the scaled incremental fuzzy PID control action ($\Delta u_s^{PID}(k)$). S is the scaling factor and a unit delay element (z^{-1}) is used to obtain the final fuzzy PID control action ($u^{PID}(k)$).

Looking at Fig. 2, one can obtain the incremental fuzzy PID control action, similar to Eq. (3), as follows:

$$\Delta u_s^{PID}(k) = \gamma_s^P \Delta e(k) + \gamma_s^I e(k) + \gamma_s^I \Delta^2 e(k) \tag{5}$$

Here, γ_s^P, γ_s^I, and γ_s^D are the gains of fuzzy proportional, fuzzy integral, and fuzzy derivative controllers, respectively. Unlike previous works [10, 18, 21], it should be noted that the gains γ_s^P, γ_s^I, and γ_s^D are only dependent on $\Delta e(k)$, $e(k)$, and $\Delta^2 e(k)$, respectively, i.e. $\gamma_s^P = f(\Delta e(k))$, $\gamma_s^I = f(e(k))$, and $\gamma_s^D = f(\Delta^2 e(k))$. This allows us to tune each of the gains independently, irrespective of the other gains. Next we discuss the principle components of an IT2 TS fuzzy controller.

2.1 Fuzzification

A typical IT2 fuzzy set [10, 19] for control application is shown in Fig. 3. The input x (e, Δe, or $\Delta^2 e$) is fuzzified by negative (N) and positive (P) fuzzy sets. The universe of discourse (UoD) of the underlying type-1 fuzzy set (shown in dotted lines in Fig. 3) is $[-l_x, l_x]$. An IT2 fuzzy set is obtained by incorporating p_x deviation to UoD of type-1 fuzzy set, i.e $[-l_x, l_x]$. Hence, the UoD of IT2 fuzzy set is $[-l_x - p_x, l_x + p_x]$. Due to the deviation p_x, the membership functions of the fuzzy sets N and P becomes uncertain. This uncertainty is captured by θ_x, given as $\theta_x = p_x / (2 \cdot l_x)$. Mathematically the fuzzy membership functions are given as follows:

$$\underline{\mu}_N(x) = \begin{cases} -\frac{1}{2 \cdot l_x} x + (0.5 - \theta_x) & \text{if } -l_x - p_x \le x \le l_x - p_x \\ 0 & \text{if } l_x - p_x \le x \le l_x + p_x \end{cases}$$

$$\bar{\mu}_N(x) = \begin{cases} 1 & \text{if } -l_x - p_x \le x \le -l_x + p_x \\ -\frac{1}{2 \cdot l_x} x + (0.5 + \theta_x) & \text{if } -l_x + p_x \le x \le l_x + p_x \end{cases}$$

$$\underline{\mu}_P(x) = \begin{cases} 0 & \text{if } -l_x - p_x \le x \le -l_x + p_x \\ \frac{1}{2 \cdot l_x} x + (0.5 - \theta_x) & \text{if } -l_x + p_x \le x \le l_x + p_x \end{cases}$$

$$\bar{\mu}_P(x) = \begin{cases} \frac{1}{2 \cdot l_x} x + (0.5 + \theta_x) & \text{if } -l_x - p_x \le x \le l_x - p_x \\ 1 & \text{if } l_x - p_x \le x \le l_x + p_x \end{cases}$$

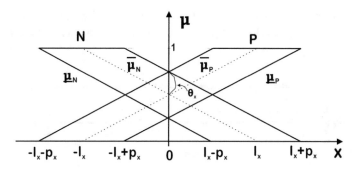

Fig. 3 Membership functions of interval type-2 fuzzy sets

2.2 TS Rule Base

The TS rule base consists of IF-THEN statements incorporating the control strategy. Here, single-input rule base for fuzzy integral control action is given as follows:

R_1^I: IF $e(k)$ is N THEN $\Delta u_1^I(k) = a_1^I e(k) + a_0^I$
R_2^I: IF $e(k)$ is P THEN $\Delta u_2^I(k) = b_1^I e(k) + b_0^I$

Similarly the control rules for fuzzy proportional and fuzzy derivative control actions are given as follows:

R_1^P: IF $\Delta e(k)$ is N THEN $\Delta u_1^P(k) = a_1^P \Delta e(k) + a_0^P$
R_2^P: IF $\Delta e(k)$ is P THEN $\Delta u_2^P(k) = b_1^P \Delta e(k) + b_0^P$

R_1^D: IF $\Delta^2 e(k)$ is N THEN $\Delta u_1^D(k) = a_1^D \Delta^2 e(k) + a_0^D$
R_2^D: IF $\Delta^2 e(k)$ is P THEN $\Delta u_2^D(k) = b_1^D \Delta^2 e(k) + b_0^D$

In the rules, N and P represents the IT2 fuzzy sets as shown in Fig. 3, and a_1^I, a_0^I, b_1^I, b_0^I are the tuneable parameters of fuzzy integral control action. Similarly, a_1^P, a_0^P, b_1^P, b_0^P, and a_1^D, a_0^D, b_1^D, b_0^D are the tuneable parameters of fuzzy proportional and fuzzy derivative control actions, respectively. It can be seen from the rule base that the need of triangular norms and co-norms is eliminated.

2.3 Input Space

An input space or state-space describes the relation between the input fuzzy sets and the fuzzy rules. Each fuzzy rule is mapped into an individual input space as shown in Figs. 4 and 5. The overall input space is obtained by merging the input space of both the rules as shown in Fig. 6. Here, the input space is 1-D since each control action takes single-input. The input space depicts certain zones (Z_1, Z_2, Z_3) depending upon the firing intervals of each rule. The lower and upper firing intervals due to both the control rules in each zone are presented in Table 1 and depicted in Figs. 4 and 5.

The input space decides the computational complexity of the fuzzy controllers. As each zone corresponds to a unique control law, the complexity increases with the

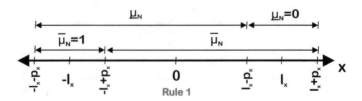

Fig. 4 1-D input space due to rule 1

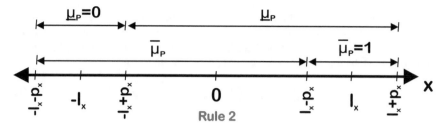

Fig. 5 1-D input space due to rule 2

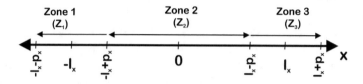

Fig. 6 1-D input space due to both rules

Table 1 Resultant membership functions due to both rules in 1-D input space

Zone	$\underline{\mu}_1$	$\bar{\mu}_1$	$\underline{\mu}_2$	$\bar{\mu}_2$
Z_1	$\underline{\mu}_N$	1	0	$\bar{\mu}_P$
Z_2	$\underline{\mu}_N$	$\bar{\mu}_N$	$\underline{\mu}_P$	$\bar{\mu}_P$
Z_3	0	$\bar{\mu}_N$	$\underline{\mu}_P$	1

increase in the number of zones. Further, with increase in the number of inputs the dimension of the input space also increases leading to complex control laws. Hence, a 1-D input space reduces the computational complexity significantly.

2.4 Type Reduction and Defuzzification

To reduce the IT2 fuzzy sets to type-1 fuzzy sets, type reduction is incorporated. This is achieved by a type reduction algorithm where the lower firing intervals of both rules are utilized to obtain the lower scaled control output ($\Delta \underline{u}_s^{CA}(k)$). Similarly the upper scaled control output ($\Delta \bar{u}_s^{CA}(k)$) is obtained from the upper firing intervals. Mathematically these are given by

$$\Delta \underline{u}_s^{CA}(k) = \frac{\underline{\mu}_1 \cdot \Delta u_1^{CA}(k) + \underline{\mu}_2 \cdot \Delta u_2^{CA}(k)}{\underline{\mu}_1 + \underline{\mu}_2}$$

$$\Delta \bar{u}_s^{CA}(k) = \frac{\bar{\mu}_1 \cdot \Delta u_1^{CA}(k) + \bar{\mu}_2 \cdot \Delta u_2^{CA}(k)}{\bar{\mu}_1 + \bar{\mu}_2}$$

(6)

where CA denotes control action which is either fuzzy proportional (P), or integral (I) or derivative (D). Each of the lower and upper scaled control outputs are treated as individual type-1 fuzzy control outputs. The final crisp output is computed by taking the average of the lower and upper scaled control outputs, i.e.,

$$\Delta u_s^{CA} = \{\Delta \underline{u}_s^{CA}(k) + \Delta \bar{u}_s^{CA}(k)\}/2 \tag{7}$$

3 Controller Structure

In this section, we discuss about the IT2 TS F-P/F-I/F-D controller structure in 1-D input space. The general structure is given in Eq. (5) where $\gamma_s^P = f(\Delta e(k))$, $\gamma_s^I = f(e(k))$, and $\gamma_s^D = f(\Delta^2 e(k))$. Here we present the explicit form of the function $f(\cdot)$. The mathematical expressions of the function $f(\cdot)$ is derived from the results obtained in Table 1 and Eqs. (6) and (7).

Considering fuzzy integral (I) action, the firings intervals $[\underline{\mu}_1, \ \bar{\mu}_1]$ (for rule 1) and $[\underline{\mu}_2, \ \bar{\mu}_2]$ (for rule 2) is obtained for zones Z_1, Z_2, and Z_3 from Table 1. Their membership grades are provided in Sect. 2.1. Using these expressions in Eq. (6), we obtain the lower and upper scaled control outputs ($\Delta \underline{u}_s^I(k)$ & $\Delta \bar{u}_s^I(k)$) for the corresponding zones as given in Eqs. (8)–(10).

Zone, Z_1

$$\Delta \underline{u}_s^I(k) = a_1^I e + a_0^I$$
$$\Delta \bar{u}_s^I(k) = \frac{(b_1^I e + b_0^I + l_e b_1^I + 2l_e a_1^I + 2l_e \theta_e b_1^I)e + 2l_e(a_0^I + 0.5b_0^I + \theta_e b_0^I)}{e + 2l_e(1.5 + \theta_e)} \tag{8}$$

Zone, Z_2

$$\Delta \underline{u}_s^I(k) = \frac{e}{4l_e(0.5 - \theta_e)}(-a_1^I e - a_0^I + b_1^I e + b_0^I) + \frac{1}{2}(a_1^I e + a_0^I + b_1^I e + b_0^I)$$
$$\Delta \bar{u}_s^I(k) = \frac{e}{4l_e(0.5 + \theta_e)}(-a_1^I e - a_0^I + b_1^I e + b_0^I) + \frac{1}{2}(a_1^I e + a_0^I + b_1^I e + b_0^I) \tag{9}$$

Zone, Z_3

$$\Delta \underline{u}_s^I(k) = b_1^I e + b_0^I$$
$$\Delta \bar{u}_s^I(k) = \frac{(-a_1^I e - a_0^I + l_e a_1^I + 2l_e b_1^I + 2l_e \theta_e a_1^I)e + 2l_e(b_0^I + 0.5a_0^I + \theta_e a_0^I)}{-e + 2l_e(1.5 + \theta_e)} \tag{10}$$

Now, using the expressions of $\Delta \underline{u}_s^I(k)$ & $\Delta \bar{u}_s^I(k)$ for zones Z_1, Z_2, and Z_3 in Eq. (7) we obtain the final crisp output Δu_s^I for fuzzy integral action. Similarly the expressions of $f(\cdot)$ is obtained for fuzzy proportional and derivative action. The expressions are similar to Eqs. (8)–(10) where $'I'$ is replaced with $'P'$ for fuzzy proportional action and with $'D'$ for fuzzy derivative action. The final fuzzy PID control action is the sum of the fuzzy proportional, integral, and derivative actions.

3.1 Properties

The properties of incremental IT2 TS F-P/F-I/F-D is enunciated below:

1. The incremental IT2 TS F-P/F-I/F-D controller is a nonlinear variable gain controller as the gains are the function of $\Delta e(k)$, $e(k)$, and $\Delta^2 e(k)$.
2. The incremental IT2 TS F-P/F-I/F-D controller becomes a incremental type-1 fuzzy P/I/D controller when $\theta_{\Delta e} = \theta_e = \theta_{\Delta^2 e} = 0$.

   a. The UoD of incremental IT2 TS F-P/F-I/F-D controller gets reduced to $[-l_{\Delta e}, l_{\Delta e}]$, $[-l_e, l_e]$, and $[-l_{\Delta^2 e}, l_{\Delta^2 e}]$ when $\theta_{\Delta e} = \theta_e = \theta_{\Delta^2 e} = 0$.
   b. Zones Z_1 and Z_3 vanish when $\theta_{\Delta e} = \theta_e = \theta_{\Delta^2 e} = 0$. Only zone Z_2 is present.

3. The control action for F-P is shown in Fig. 7 where the values are chosen as $a_1^P = 5$, $a_0^P = 0.1$, $b_1^P = 3$, $b_0^P = 0.2$, $l_{\Delta e} = 1$, and $\theta_{\Delta e} = 0.125$ without loss of generality.
4. The control action generated by F-P is $(a_0^P + b_0^P)/2$ at $\Delta e(k) = 0$. This is due to the offset term present in the control rules. Same reasoning is applicable to F-I and F-D control actions.
5. The control action generated by the incremental IT2 TS F-P/F-I/F-D controller is continuous in nature and it increases monotonically from minimum to the maximum value as seen from Fig. 7.

Remark The proposed IT2 TS fuzzy PID controller is obtained by individually combining the incremental F-P, F-I, and F-D control actions. However, the control effort becomes zero at $e(k) = \Delta e(k) = \Delta^2 e(k) = 0$ only if the offset terms (a_0, b_0) become zero.

Fig. 7 Control action due to F-P

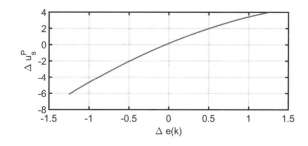

4 Simulation Study

In this section, we employ the IT2 fuzzy P/I/D controller on some unstable/nonlinear systems to evaluate the performance of the proposed scheme. Furthermore we compare the results of the proposed method with linear PI, variable structure controller (VSC) [25], and variable structure fuzzy controller (VSFC) [26].

Example 1 An unstable system with dead-time is considered which is given by the transfer function $F(s)$ as follows:

$$F(s) = e^{-0.25s}/(s - 1) \tag{11}$$

These type of systems are quite common in process control industries. Here, the output of the system is to be regulated to a constant value. For this, the controller parameters are tuned as given below:

IT2 Fuzzy P: $a_1^P = 2$, $a_0^P = 30$, $b_1^P = 7$, $b_0^P = 60$, $l_{\Delta e} = 1$.
IT2 Fuzzy I: $a_1^I = 0.0016$, $a_0^I = 0.024$, $b_1^I = 0.0056$, $b_0^I = 0.048$, $l_e = 1250$.
IT2 Fuzzy D: $a_1^D = 120$, $a_0^D = 1800$, $b_1^D = 420$, $b_0^D = 3600$, $l_{\Delta^2 e} = 0.017$, and $S = 1.5$.

The parameters of linear PI, VSC, and VSFC controllers are provided in [26].

The set point responses are presented in Fig. 8a. To compare the performances, we present a few performance metrics such as integral square error (ISE), integral time absolute error (ITAE), rise time (t_r), percentage overshoot ($\%M_p$), and settling time (t_s). All these values are tabulated in Table 2 for simulation time t = 10 s. Furthermore, a step disturbance is introduced at 25 s to check the robustness of the proposed scheme. The output responses in presence of disturbances are shown in Fig. 8b. The proposed method performs significantly better in terms of ITAE, ISE, t_r, t_s, and disturbance rejection in comparison to its counterparts.

Example 2 A continuous stirred tank reactor (CSTR) system with three state variables [25, 27] is considered. The dynamics of the highly nonlinear CSTR system is given as follows:

$$\begin{aligned}
\dot{x}_1 &= -(1 + \delta_1)x_1 + 2\delta_2 x_{2d} x_2 + \delta_2 x_2^2 \\
\dot{x}_2 &= \delta_1 x_1 - (1 + \delta_2 x_{2d} + 2\delta_3 x_{2d})x_2 - (\delta_2 + \delta_3)x_2^2 + u \\
\dot{x}_3 &= 2\delta_3 x_{2d} x_2 - x_3 + \delta_3 x_2^2
\end{aligned} \tag{12}$$

where x_1, x_2, and x_3 are the concentrations (dimensionless) of the reactants in the reactor zone, x_2 is the measurable reactant concentration in the exit stream, and the manipulated variable u is the feed concentration (dimensionless). The system parameters are given by $\delta_1 = 3$, $\delta_2 = 0.5$, $\delta_3 = 1$, and $x_{2d} = 0.8796$.

CSTR is a benchmark and challenging problem of process control industry. Here, the feed concentration x_2 is to be regulated to a constant value. The controllers are tuned to maintain the feed concentration and the tuned parameters are as follows:

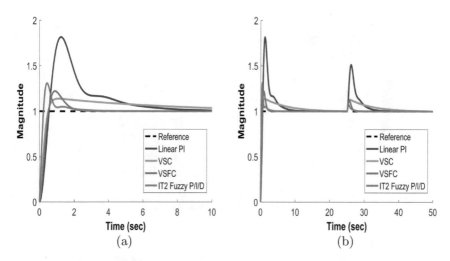

Fig. 8 Set-point responses (Example 1) **a** normal condition and **b** in the presence of input disturbances at t = 25 s

Table 2 Performance metrics (Example 1)

Controller	ITAE	ISE	t_r (in s)	% M_p	t_s (in s)
Linear PI	3.985	0.9615	0.56	81%	8.1
VSC [25]	3.224	0.2162	0.51	14%	13.5
VSFC [26]	0.2726	0.2305	0.53	22%	2.2
IT2 fuzzy P/I/D (proposed)	0.2143	0.1063	0.22	31%	2.1

IT2 Fuzzy P: $a_1^P = 2.4$, $a_0^P = 36$, $b_1^P = 8.4$, $b_0^P = 72$, $l_{\Delta e} = 0.83$.
IT2 Fuzzy I: $a_1^I = 0.007$, $a_0^I = 0.105$, $b_1^I = 0.0245$, $b_0^I = 0.21$, $l_e = 28.57$.
IT2 Fuzzy D: $a_1^D = 12$, $a_0^D = 180$, $b_1^D = 42$, $b_0^D = 360$, $l_{\Delta^2 e} = 1.67$, and $S = 2$.
The parameters of linear PI, VSC, and VSFC controllers are provided in [26].

The output responses with linear PI, VSC, VSFC, and IT2 fuzzy P/I/D controllers are shown in Fig. 9a. Here, a step disturbance is introduced at 3 s to check the robustness. Further, tracking performance of the controllers is investigated and the obtained responses are shown in Fig. 9b. The performance metrics ITAE, ISE, t_r, %M_p, and t_s for the four controllers are tabulated in Table 3 for simulation time t = 5 s. It can be deduced from Fig. 9 and Table 3 that IT2 fuzzy P/I/D controller performs satisfactorily.

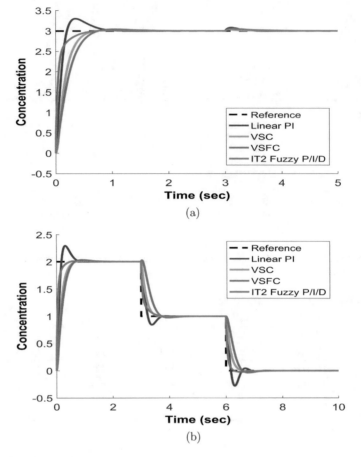

Fig. 9 CSTR (Example 2) **a** set-point responses in the presence of input disturbance at t = 3 s, and **b** tracking responses

Table 3 Performance metrics (Example 2)

Controller	ITAE	ISE	t_r (in s)	% M_p	t_s (in s)
Linear PI	0.1415	0.493	0.2	10%	0.8
VSC [25]	0.1086	0.9303	0.68	0.4%	0.75
VSFC [26]	0.2213	1.097	0.7	1.3%	0.75
IT2 fuzzy P/I/D (proposed)	0.1446	0.2124	0.6	0.5%	0.7

5 Discussions and Conclusions

In this chapter we have presented a simplified structure of IT2 TS fuzzy PID controller using a 1-D input space (state-space). The fuzzy PID control action is obtained by individually modeling and combining F-P, F-I and F-D control actions in 1-D input space. The computational complexity of control laws drastically reduces as the need of triangular norms and co-norms is eliminated in 1-D input space. In this approach the input space is divided into three zones where a distinct control law is activated in each zone. The control laws obtained are validated on two systems, unstable first-order system with time delay and CSTR. Both these systems are quite prevalent in the process control industries. The results obtained with the proposed IT2 fuzzy PID controller are compared with linear PI, VSC, and VSFC controllers. It is observed that proposed scheme performs significantly better than its counterparts.

The approach in this chapter uses an integer order fuzzy controller. However, it can be extended to a fractional order fuzzy controller. Fractional order fuzzy PID controller introduces two extra degree of freedom which can provide further flexibility in controller tuning. However, fuzzy PID controllers generally have a large number of parameters to tune. To simplify the tuning process, one can employ any heuristic based optimization algorithm such as genetic algorithm, particle swarm optimization etc to obtain optimal performances.

References

1. Zadeh, L.A.: Outline of a new approach to the analysis of complex systems and decision processes. IEEE Trans. Syst., Man, Cybernet. **SMC-3**(1), 28–44 (1973)
2. Mamdani, E.H.: Application of fuzzy algorithms for control of simple dynamic plant. Proc. Inst. Electr. Eng. **121**(12), 1585–1588 (1974)
3. Ying, H.: The simplest fuzzy controllers using different inference methods are different nonlinear proportional-integral controllers with variable gains. Automatica **29**(6), 1579–1589 (1993)
4. Mizumoto, M.: Realization of PID controls by fuzzy control methods. Fuzzy Sets Syst. **70**(2–3), 171–182 (1995)
5. Misir, D., Malki, H.A., Chen, G.: Design and analysis of a fuzzy proportional-integral-derivative controller. Fuzzy Sets Syst. **79**(3), 297–314 (1996)
6. Mann, G.K.I., Hu, B.G., Gosine, R.G.: Analysis of direct action fuzzy PID controller structures. IEEE Trans. Syst., Man, Cybern. Part B **29**(3), 371–388 (1999)
7. Golob, M.: Decomposed fuzzy proportional-integral-derivative controllers. Appl. Soft Comput. **1**(3), 201–214 (2001)
8. Raj, R., Mohan, B.M.: Modeling and analysis of the simplest fuzzy PID controller of Takagi-Sugeno type with modified rule base. Soft. Comput. **22**(15), 5147–5161 (2018)
9. Sain, D., Mohan, B.M.: Modelling of a nonlinear fuzzy three-input PID controller and its simulation and experimental realization. IETE Tech. Rev. **38**(5), 479–498 (2021)
10. Raj, R., Mohan, B.M., Lee, D.E., Yang, J.M.: Derivation and structural analysis of a three-input interval type-2 TS fuzzy PID controller. Soft Comput. **26**, 589–603 (2022)
11. Sain, D., Mohan, B.M.: A simple approach to mathematical modelling of integer order and fractional order fuzzy PID controllers using one-dimensional input space and their experimental realization. J. Franklin Inst. **358**(7), 3726–3756 (2021)

12. Das, S., Pan, I., Das, S., Gupta, A.: A novel fractional order fuzzy PID controller and its optimal time domain tuning based on integral performance indices. Eng. Appl. Artif. Intell. **25**(2), 430–442 (2012)
13. Kumar, V., Rana, K.P.S., Kumar, J., Mishra, P., Nair, S.S.: A robust fractional order fuzzy P+ fuzzy I+ fuzzy D controller for nonlinear and uncertain system. Int. J. Autom. Comput. **14**(4), 474–488 (2017)
14. Kumar, A., Raj, R.: Design of a fractional order two layer fuzzy logic controller for drug delivery to regulate blood pressure. Biomed. Signal Process. Control **78**, 104024 (2022)
15. Raj, R., Mohan, B.M.: Stability analysis of general Takagi-Sugeno fuzzy two-term controllers. Fuzzy Inf. Eng. **10**(2), 196–212 (2018)
16. Raj, R., Mohan, B.M.: Analytical structures and stability analysis of the simplest Takagi-Sugeno fuzzy two-term controllers. Int. J. Process Syst. Eng. **5**(1), 67–92 (2019)
17. Zhou, H., Ying, H., Zhang, C.: Effects of increasing the footprints of uncertainty on analytical structure of the classes of interval type-2 Mamdani and TS fuzzy controllers. IEEE Trans. Fuzzy Syst. **27**(9), 1881–1890 (2019)
18. Raj, R., Mohan, B.M.: General structure of interval type-2 fuzzy PI/PD controller of Takagi-Sugeno type. Eng. Appl. Artif. Intell. **87**, 103273 (2020)
19. Raj, R., Mohan, B.M., Yang, J.M.: A simplified structure of the simplest interval type-2 fuzzy two-term controller. IFAC-PapersOnLine **53**(1), 661–666 (2020)
20. Kumari, K., Mohan, B.M.: Minimum t-norm leads to unrealizable fuzzy PID controllers. Inf. Sci. (2021)
21. Zhou, H., Zhang, C., Tan, S., Dai, Y., Duan, J.A.: Design of the footprints of uncertainty for a class of typical interval type-2 fuzzy PI and PD controllers. ISA Trans. **108**, 1–9 (2021)
22. Mendel, J.M.: Explaining the performance potential of rule-based fuzzy systems as a greater sculpting of the state space. IEEE Trans. Fuzzy Syst. **26**(4), 2362–2373 (2017)
23. Bhattacharya, S., Chatterjee, A., Munshi, S.: An improved PID-type fuzzy controller employing individual fuzzy P, fuzzy I and fuzzy D controllers. Trans. Inst. Meas. Control. **25**(4), 352–372 (2003)
24. Kumar, V., Mittal, A.P.: Parallel fuzzy P+ fuzzy I+ fuzzy D controller: design and performance evaluation. Int. J. Autom. Comput. **7**(4), 463–471 (2010)
25. Ablay, G.: Variable structure controllers for unstable processes. J. Process Control **32**, 10–15 (2015)
26. Raj, R.: Variable structure fuzzy controller for unstable processes. In: Soft Computing: Theories and Applications, pp. 719–728. Springer, Singapore (2022)
27. Ghaffari, V., Naghavi, S.V., Safavi, A.A.: Robust model predictive control of a class of uncertain nonlinear systems with application to typical CSTR problems. J. Process Control **23**(4), 493–499 (2013)

Fractional Order Interval Type-2 Fuzzy Logic Controller

Snehanshu Shekhar and Anupam Kumar

Abstract This chapter gives an elucidation of fractional order calculus, its extension in Fractional Order PID, fractional order Fuzzy type-1 and interval type-2 controllers and their applications. Further, the proposed IT2FOFPID controller is employed to control fractional order plants. The Influence of fractional order λ and μ on the IT2FOFPID controller is studied in details. Further in order to the shows the superiority of the proposed controller, the performance comparison of the proposed controllers for step response is also done for fractional order plants. At last from the simulation results, it is clear that IT2FOFPID controller is giving better results in terms of ITAE, overshoot etc with existing controllers.

1 Introduction

Fractional calculus, characterised by differentiation or integration of non-integer order, is recognised as an efficient tool to represent real dynamic systems. The concept of fractional calculus dates back to 3rd century B.C., Archimedes introduced in his book, "*The Method of Mechanical Theorems*", popular as "*The Method*". The details of preliminary calculus even is found in the book, *Archimedes Palimpsest* [2]. The concept of power was pioneered in its modern form by René Descartes. The work of Fermat and Pascal into the calculus of probabilities laid important platform for Leibniz' formulation of the infinitesimal/fractional calculus which resulted controversy between Newton and Leibniz (the Leibniz-Newton calculus controversy). The term "Fractional calculus" was originally coined to justify the extended implementation of a derivative of integer order $\frac{d^n y}{dx^n}$ to define a fraction. Later the question

S. Shekhar
Electronics and Communication Engineering, Indian Institute of Information Technology Kota, Jaipur, India

A. Kumar (✉)
Main Supervisor, Electronics and Communication Engineering, National Institute of Technology Patna, Patna, India
e-mail: anuanu1616@gmail.com; anupam.ec@nitp.ac.in

© The Author(s), under exclusive license to Springer Nature Switzerland AG 2023
O. Castillo and A. Kumar (eds.), *Recent Trends on Type-2 Fuzzy Logic Systems: Theory, Methodology and Applications*, Studies in Fuzziness and Soft Computing 425,
https://doi.org/10.1007/978-3-031-26332-3_3

became: "Can n be any number- fractional, irrational or complex?" Due to this query, the name fractional calculus is preferable known as integration/differentiation of any order [3]. Ever since the invention of notation $\frac{d^n y}{dx^n}$ by Leibniz, there has been a major shift in solving the real system having uncertainty as its inherent part which affects decision-making component of intelligent system. Leibniz's "*general order derivative*" definition was used by Johann Bernoulli, John Wallis (1697) and later by Leonhard Euler to express derivative algebraically (1730 and 1738). In 1772, the famous Lagrange further extended the work of Leibniz to develop the law of exponents to solve differential operators of integer order as:

$$\frac{d^m}{dx^m} \cdot \frac{d^n}{dx^n} y = \frac{d^{m+n}}{dx^{m+n}} y \tag{1}$$

The *dot* has been eluded in modern notation. A French mathematician and astronomer, Laplace, extended the works done by his predecessors like Adrien-Marie Legendre (Legendre function and symbol Γ). Pierre-Simon Laplace in his five-volume Mécanique Céleste (Celestial Mechanics) (1799–1825). It included the translation of the conventional geometric techniques to calculus and defining fractional derivative by an integral (1820 vol. 3, 85 and 186). Works of Laplace including Laplace transform, application of calculus to potential function, astro-physics and mechanics, theory of probabilities and many more changed the perception of calculus and gave it a novel dimension of application. Euler and Lagrange were pioneer to contribute in this field in the 1800s, while Abel was the first to implement it. The first categorized works were done by the mid of 20th century by Liouville, Riemann and Holmgren. The nth-order series and its expansion of exponentials was defined by Liouville. Riemann contributed in solving for a definite integral. However, the deductions of Liouville and Riemann [23, 24] were unified by Grunwald and Krug.

The requirement for solving the feedback amplifier was the vital step towards implementation of this fractional calculus by engineers. Solution proposed by Bode for a feedback loop, the performance could elude the effect of the changes in the gain of the closed-loop amplifier [1] and proved to be game changer for application in engineering. The solution, primarily, coined as "ideal cut-off characteristic" and currently known as "Bode's ideal loop transfer function" greatly abet in increasing the robustness of the system to parameter changes, uncertainties or disturbances. The works of Bode was carried forward fractional calculus systems, which consists of both theory and applications. Recently, the practical applications of fractional calculus witnessed a growth in many field like feedback control, signal processing and many more.

Dynamic models of most of the real systems are represented using integer-order differentiation and integration-based traditional calculus. Practically, these dynamic systems are constituted using real objects, generally, observing behaviour that are fractional in nature. However, for many of them, the degree of fractionality is very low [1, 4, 5]. The foundation of PID controller is marked by the work of Taylor and Foxboro (1930) who installed 'turn key' control systems. This embedded the derivative action [35, 36]. This initial work is the base of current trends of controllers like

Tilt-Integral-Derivative (TID) controller, multiterm fractional order PID (MFOPID) controller and complex fractional order PID (CFOPID) controller [25]. The conventional PID controller which is based on terms such as D-term and I-term that can be generalized into non-integer order operations. The obtained controller is famously known as Fractional order controller (FOC). FOC implementation on real system has become possible only because of oustaloup recursive approximation (ORA) method [1, 5, 7, 28, 33]. Use of FOC has been extended to fractional order fuzzy logic (FOFPID) controller as a result of merging fractional order controller with fuzzy logic controller [26]. Evolution of type-1 and type-2 Fractional Order Fuzzy Logic Controllers was a leap step for controlling complex systems [10, 13, 16, 17]. Type-1 FOFPID (T1FOFPID) controller finds application in a variety of field including electrical systems [11, 12] to abreast field of biomedical [15]. In 2017, enhancement of IT2FLC with fractional order known as the interval type-2 fractional order fuzzy logic PID (IT2FOFPID) controllers were first time proposed by Kumar and Kumar [9] and given new direction for research community [29]. After that many papers have been reported on this concept. It shows that the proposed techniques are well accepted among the researchers [38, 39]. Further, augmentation of result and controlling through IT2FOFPID controller has resulted in its application in versatile fields including economics, engineering, sports, sociology, finance, etc. [19].

2 Interval Type-2 Fuzzy Logic Controller

2.1 Oustaloup Recursive Approximation

Implementation of fractional controller requires thorough understanding in design and implementation of robust and practical controller. It is generally implemented with proper approximation of the fractional integrals and derivatives with rational transfer functions. This approximation implements Oustaloup Recursive Approximation method [1, 7]. A general fractional derivative of order γ can be defined by an Eq. (2) as:

$$G_f(s) = s^\gamma \tag{2}$$

Generalized fractional order derivative as represented in Eq. (2) can be approximated with rational transfer function, as shown in Eq. (3):

$$G_f(s) = s^\gamma = K \prod_{k=-N}^{k=N} \frac{(s + w_k')}{(s + w_k)} \tag{3}$$

The poles (w_k), zeros (w_k'), and gain (K) is calculated with the Eq. (4)

$$w_k = w_l \left(\frac{w_h}{w_l}\right)^{\frac{K+N+\frac{1}{2}(1+\gamma)}{2N+1}}, (w_k)' = w_l \left(\frac{w_h}{w_l}\right)^{\frac{K+N+\frac{1}{2}(1-\gamma)}{2N+1}}, \text{ and } K = (w_k') \tag{4}$$

Fig. 1 Oustaloup recursive approximation block diagram [21]

where, w_h and w_l are ranges of higher and lower frequencies and $2N + 1$ is the order of the system. It can be noted that Eq. (3) can be also represented using block diagram as shown in Fig. 1.

2.2 Fractional Order PID Controller

Basic control action of controller in a feedback system includes proportional, derivative and integral controls. History of PID controller dates back to 1907 when Tagliabue Company developed the first automated temperature control system [37]. Its effects on controlled system behaviour include:

- For proportional control: The output response of controlled plant increases, steady-state error response and relative stability decrease.
- For derivative control: The over all relative stability along with sensitivity to noise of the system increases.
- For integral control: The steady-state error of the system eliminates however decreases the relative stability.

Fractional order controller finds application in a diversified field due to which numerous works have been done for tuning the control parameters for better performances. Tuning method includes biggest log-modulus tuning (BLT) method and internal model control (IMC) method of designing conventional PID controllers, amalgamation of premier methods, Cross-Entropy Method (CEM), for tuning controllers for multi-variable processes [28, 33]. The researcher Podlubny was the first to present a generalised form of PID control as $PI^\lambda D^\mu$ where integrator and the differentiator have the order of λ and μ, respectively.

Additionally, he depicted performances of fractional-order PID and integer order PID for controlling the fractional order plants [5]. PID and FOPID controllers are special forms of controllers represented by the generalized form, as shown in Fig. 3. The pictorial representation of a generalized PID control systems are demonstrated in Fig. 2.

In time domain, FOPID controller is represented by a generalized equation (Eq. 5):

$$u(t) = K_P e(t) + K_I I_t^{-\lambda} e(t) + K_D D_t^\mu e(t) \tag{5}$$

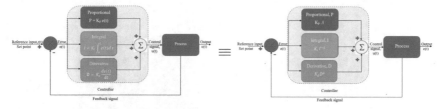

Fig. 2 Fractional order PID representation

Equation (5) can also be represented as

$$u(t) = K_P e(t) + K_I \frac{dt^{-\lambda} e(t)}{dt^{-\lambda}} + K_D \frac{dt^{\mu} e(t)}{dt^{\mu}} \qquad (6)$$

The transfer function of FOPID controller (G(s)), in s-domain, is represented by equation:

$$G_c(s) = \frac{U(s)}{E(s)} = K_P + K_I \frac{1}{s^{\lambda}} + K_D s^{\mu}, \qquad (7)$$

Equations (5), (6) and (7) are formulated with the following conditions:

1. $\lambda, \mu > 0$
2. System is considered to be at 0 initial conditions.

Fractional order PID controller contains two additional parameters (λ, μ) in comparison to classical controller (K_P, K_I and K_D) for tuning which adds more control reliability to the system [6, 7]. Important advantages include augmented closed-loop performance, better noise/disturbance rejection capability, surged control of time-delay system and increased robustness. The integration function eliminates the static error, eluding error tracking of controlled variables to reference/set-point value, and increases accuracy of control of systems. The degree of integration effect is dependent on the integral coefficient K_I which in turn in dependent upon λ. The derivative parameter, K_D improves the dynamic characteristics, without effecting steady-state process. The variation trend of deviation signal (error) is echoed by differentiation element, and an effective modified initial signal can be brought into the system before the oversize of the deviation signal value. Overshoot, system noise and disturbance rejection ability and rise time are inversely proportional to K_D. The step responses of system for different values of μ and λ can be seen in the case study section.

In addition, the FOPID control techniques are the classical PID controller and expands it from point to plane for different controllers as described below and shown in Fig. 3.

- P controller: When $\lambda = 0$ and $\mu = 0$,
- PI controller: When $\lambda = 1$ and $\mu = 0$,
- PD controller: When $\lambda = 0$ and $\mu = 1$,
- PID controller: When $\lambda = 1$ and $\mu = 1$.

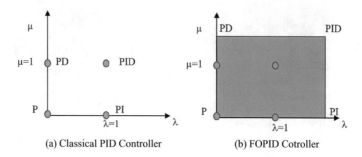

(a) Classical PID Controller   (b) FOPID Cotroller

Fig. 3 Representation of conventional fractional order in two dimension

2.3 Interval Type-2 Fractional Order Fuzzy PD+I Controller

An IT2FOFPID controller is a controller implementing type-2 FLC with fractional order enhancement. A type-2 fuzzy set is characterized by a three-dimensional membership function. Type-2 fuzzy logic is motivated from the fact of ambiguity in perseverance of degree of belonging or membership function. The membership function of a type-2 fuzzy set is three dimensional.

In addition, basic of all types of fuzzy system (*type-n*) remain the same [20]. However, "degree of fuzziness" is proportional to the type number. Higher type changes the nature of the membership functions, thereby changing the operations depending on the membership functions. The basic structure of the type-2 fuzzy rules is similar to that of type-1 because the distinction between both the types is associated with the nature of the membership functions. Hence, the only variation is that in the sets involved in the rules of the types. In type-1 fuzzy system, defuzzification is done to obtain a crisp output which may be seen as crisp (type-0) representative of the combined output sets. In type-2 fuzzy system, output being type-2, an extended versions of type-1 defuzzification is used since defuzzification operation in the type-2 gives a type-1 fuzzy set at the output. As type is reduced to type-1 due to defuzzification, this operation is called "type reduction" and the type-1 fuzzy set so obtained is called "type-reduced set". The type-reduced fuzzy set is further be defuzzified to obtain a single crisp number, depending upon the applications and requirement. A generalized fuzzy type-2 system is depicted in Fig. 4. An additional stage, type reducer implementing Type-Reduction algorithm, abets the output processor for the conversion of type-2 FS output into an equivalent type-1 FS [20]. For control applications, the generalized closed-loop feedback FLC is depicted in the Fig. 5 for controlling the plants.

Furthermore, the mathematical analysis of proposed controller is demonstrated. Therefore, the output of IT2FOFPD+I controller (referred here as *IT2FOFPID* controller) can be determined considering each of the gains individually and summing them [29, 32]. The output of proportional control action is given as:

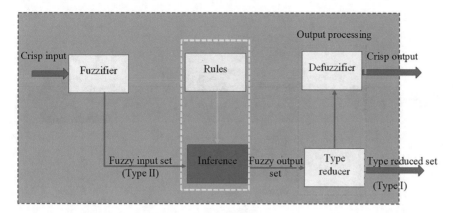

Fig. 4 Type-2 fuzzy logic system

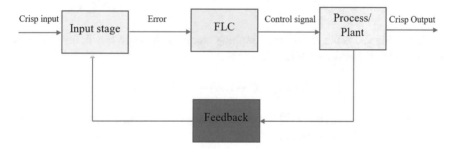

Fig. 5 Generalized closed-loop feedback FLC system

$$E = G_P * e(t) \tag{8}$$

where G_P is the gain of the proportional controller.

The output of derivative control action is given as,

$$CE = G_{De} * \frac{d^\mu e(t)}{dt^\mu} \tag{9}$$

where G_{De} is the gain of the derivative controller.

Similarly output of integral control action is given as:

$$IE = G_{Ie} * \frac{d^{-\lambda} e(t)}{dt^{-\lambda}} \tag{10}$$

where G_{Ie} is the gain of the integral controller.

Output of fractional order fuzzy PD controller is given as:

$$f(E + CE) \tag{11}$$

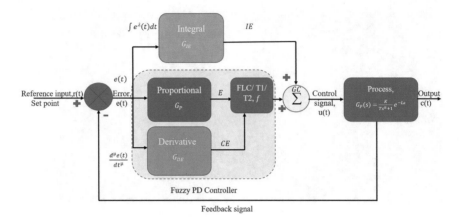

Fig. 6 IT2FOFPD+I/T1FOFPD+I controller

Assuming GC is the gain of the summing block, as shown in Fig. 6, the output of the overall combined controller is given as,

$$u(t) = [f(E + CE) + IE] * GC \tag{12}$$

Combining Eqs. (6) and (12) and approximating the non-linearity of control surface of PD rule base (*function, f*) we get,

$$u(t) \approx \left[G_P * e(t) + G_{De} * \frac{d^\mu e(t)}{dt^\mu} + G_{Ie} * \frac{d^{-\lambda} e(t)}{dt^{-\lambda}} \right] * GC \tag{13}$$

$$u(t) = G_P * GC \left[e(t) + \frac{G_{De}}{G_P} * \frac{d^\mu e(t)}{dt^\mu} + \frac{G_{Ie}}{G_P} * \frac{d^{-\lambda} e(t)}{dt^{-\lambda}} \right] \tag{14}$$

On comparing Eqs. (6) and (14) at the end, we can conclude that the scaling factors proposed controllers can be represented in terms of fractional order PID gains,

$$G_P * GC = K_P \tag{15}$$

$$\frac{G_{De}}{G_P} = K_D \tag{16}$$

$$\frac{G_{Ie}}{G_P} = K_I \tag{17}$$

Here, the five parameters of proposed controller such as G_P, GC, G_{De}, G_{Ie}, μ and λ are tuning parameters for the proposed controller. In order to do fair the comparision, T1FOFPD+I controller can be easily obtained by replacing type-2 to type-1 in the nonlinear block (function f) as shown in the Fig. 6. Here, T1FOFPD+I controller combines FOPID control and type-1 FLC to obtain a robust and effective controller. Amalgamating these two techniques result in an improved controller called T1FOFPID controller [26]. T1FOFPID minimizes error component of fractional order system and the ambiguity or uncertainty of the response is handled by FLC (refer Fig. 5). A T1FPID controller is a fuzzified PID controller which acts on the same input signal, except for the control strategy being formulated as fuzzy rules. The most general fuzzy controllers include fuzzy P controller, fuzzy PD controller, fuzzy PD+I controller and fuzzy incremental controller.

3 Case Study

This section demonstrated detail case study of the controller (fractional order integro-differential operators to the IT2FOFPID controller) for fractional order plants.

3.1 Influence of Fractional Order λ and μ on the IT2FOFPD+I Controller

This subsection presents an appropriate description for addition of extra fractional order integro-differential operators to the T2FOFPID controller. On adding these extra degrees of freedom, the control performance of fractional order plants will change in some way. For this study, the considered plant is $G_P(s)$:

$$G_P(s) = \frac{1}{0.8s^{2.2} + 0.5s^{2.2} + 1} \tag{18}$$

The ABC optimization technique is used to search optimal control parameters of the proposed controller. i.e, $G_E = 1.1773$, $G_{CE} = 0.9647$, $G_{IE} = 0.4517$, $G_U = 5.6070$, $\lambda = 0.8499$, $\mu = 0.7830$.

3.1.1 The Influence of Fractional Order Integration λ

The graphical representations of step response for plant, $G_P(s)$ for $G_E = 1.1773$, $G_{CE} = 0.9647$, $G_{IE} = 0.4517$, $G_U = 5.6070$, $\mu = 0.7830$ and variations in λ from 0.1 to 1.7 with 0.2 increments i.e., $\lambda = 0.1:0.2: 1.7$, are shown in Fig. 7. From Fig. 7, it is inferred that the overshoots of system become larger, oscillation time increased, settling time elongated and steady state error of the system is also increased for undersized values of λ. When $\lambda > 1$, i.e., for larger values of λ, the output response of the system has larger steady state errors and has larger settling time. It can be concluded that the λ control parameter is not selected to undersized or oversized.

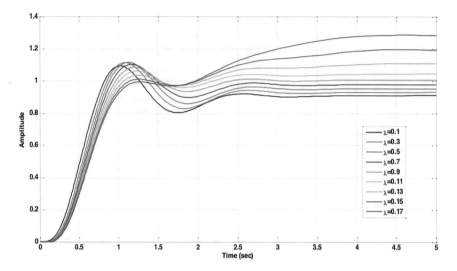

Fig. 7 The step responses of FO plant with variation in λ

The undersized value can produce larger steady state error and higher overshoots. The oversized value may lead to a higher settling time and even make the system unstable.

3.1.2 The Influence of Fractional Order Derivative μ

The step responses of fractional order plant, $G_P(s)$ for $G_E = 1.1773$, $G_{CE} = 0.9647$, $G_{IE} = 0.4517$, $G_U = 5.6070$, $\lambda = 0.8499$ and $\mu = 0.5: 0.1: 1.3$, are depicted in Fig. 7. When μ varies from 0.5 to 0.9, the overshoots and oscillation time become lower and settling time also reduced. In continuation, when μ varies from 1.0 to 1.3, the overshoot of the system becomes larger. It means that the system performance will decline for undersized or overlarge values of μ. Hence, the overlarge and undersized values of μ will cause the system unstable.

From the above study, it can be concluded that the integro-differentiation order of the TIIFOFPID controller is necessary to keep in a proper range in order to maintain in a suitable control performance (Fig. 8).

3.2 Performance Comparison of the Proposed Controller for Fractional Order Plant

This subsection presents an appropriate description for addition of additional fractional order integro-differential operators to IT2FPID controller. Annexing these additional degrees of freedom, the output performance of the proposed controller is

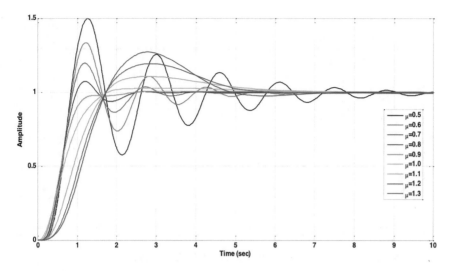

Fig. 8 The step responses of FO plant with variation in μ

studied for fractional order plant. Here, structural design of proposed IT2FOFPID controller is also shown in Fig. 6.

The non-integer order plus time delay plant, considered as case study here, is given as:

$$G_P(s) = \frac{K}{Ts^\alpha + 1} e^{-Ls} \tag{19}$$

where, $G_P(s)$, K, T, L and α are the transfer function, process gain, time constant, dead time and fractional orders of the system respectively. Here, the parameters' values are considered as $K = 0.9931$, $T = 2.3298$, $L = 1.0006$ and $= 1.0648$.

In order to present the effectiveness of the proposed controllers, the step response have been examined and compared with existing controllers such as T1FOFPD+I and PID. The objective function for optimization for this study is the ITAE. Here, the ABC optimization techniques are used for finding optimal scaling factors of the proposed controllers. The obtained optimal ITAE values for T2FOFPD+I, T1FOFPD+I and PID controllers are 3.263, 4.864 and 6.70 respectively. All the simulations study is carried out in the MATLAB.

Furthermore, the closed loop performances for proposed controllers are shown in the Fig. 9. From Fig. 9, it is clear that the unit step response of the T2FOFPD+I controller is better than that of the T1FOFPD+I and PID controllers. The proposed controller is giving similar better settling time with PID, while approximate similar settling time for T1FOFPD+I controller. The proposed controller is also showing better oscillation in response. From ITAE values, it is evident that ITAE is minimum for T2FOFPD+I controller compared with conventional counterparts. Therefore, it is clear that enhancement of the proposed controller with fractional calculus is giving better results. It can be used for other non-linear plants for better accuracy.

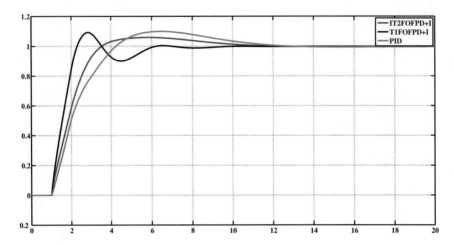

Fig. 9 The unit step responses for IT2FPD+I, T1FPD+I, PID controllers

3.3 Conclusions

In this chapter, the new combined IT2FOFPD+I controller is presented and applied successfully to control fractional order plants. The proposed controller is enhanced version of the IT2FLC to fractional order controllers. The influence of changing fractional order parameters for proposed controllers are also studied. Further, the performance comparison is also investigated for the fractional order plants. From the simulation results, it is observed that IT2FOFPD+I controller is giving better results than T1FOFPD+I and PID controller. In future, the hardware implementation needs to be done for the proposed controller which is more prone for uncertainties.

References

1. Monje, C.A., et al.: Fractional-order Systems and Controls. Fundamentals and Applications. Springer, London (2010)
2. Bell, E.T.: Men of Mathematics. Simon and Schuster (1937)
3. Ross, B.: The development of fractional calculus 1695–1900. Hist. Math. **4**(1), 75–89 (1977)
4. Petras, I., Dorcak, L.: Fractional-order control systems: modelling and simulation, fractional calculus and applied analysis. Int. J. Theory Appl. **6**(2), 205–232 (2003)
5. Podlubny, I.: Fractional-order systems and $PI^{\lambda}D^{\mu}$ -controllers. IEEE Trans. Autom. Control **44**(1), 208–214 (1999)
6. Marzak M. H. et al.: Real time performance comparison between PID and Fractional order PID controller in SMISD plant. In: IEEE 6th Control and System Graduate Research Colloqum, 2015, pp. 141–145 https://doi.org/10.1109/ICSGRC.2015.7412481
7. Baranowski, J., et al.: Time-domain oustaloup approximation. In: 20th International Conference on Methods and Models in Automation and Robotics (MMAR), 2015, pp. 116–120. https://doi.org/10.1109/MMAR.2015.7283857

8. Zadeh, L.A.: Fuzzy sets. Inf. Control **8**, 338–353 (1965). https://doi.org/10.1016/S0019-9958(65)90241-X
9. Kumar, A., Kumar, V.: A novel interval type-2 fractional order fuzzy PID controller: design, performance evaluation, and its optimal time domain tuning. In: ISA Transactions, 2017, vol. 68, pp. 251–275. ISSN 0019-0578. https://doi.org/10.1016/j.isatra.2017.03.022
10. Das, S., et al.: A novel fractional order fuzzy PID controller and its optimal time domain tuning based on integral performance indices. Eng. Appl. Artif. Intell. **25**, 430–442 (2012). https://doi.org/10.1016/j.engappai.2011.10.004
11. Tang, S., et al.: An enhanced MPPT method combining fractional-order and fuzzy logic control. IEEE J. Photovolt. **7**(2), 640–650 (2017). https://doi.org/10.1109/JPHOTOV.2017.2649600
12. Kanagaraj, N., et al.: A variable fractional order fuzzy logic control based MPPT technique for improving energy conversion efficiency of thermoelectric power generator, MDPI, 2020. Energies **13**(17):4531. https://doi.org/10.3390/en13174531
13. Liu, L., et al.: Variable-order fuzzy fractional PID controller. ISA Trans. **55**, 227–233 (2015). https://doi.org/10.1016/j.isatra.2014.09.012
14. Ghamari, S.M. et al.: Fractional-order fuzzy PID controller design on buck converter with antlion optimization algorithm. In: IET Control Theory Applications, pp. 340–352. Wiley Publisher (2022). https://doi.org/10.1049/cth2.12230
15. Kumar, A., Raj, R.: Design of a fractional order two layer fuzzy logic controller for drug delivery to regulate blood pressure. Biomed. Signal Process. Control **78**, 104024 (2022). https://doi.org/10.1016/j.bspc.2022.104024
16. Kurucu, M.C., et al.: Investigation of the Effects of Fractional and Integer Order Fuzzy Logic PID Controllers on System Performances, pp. 775–779. IEEE Xplore, ELECO (2018)
17. Belkhier, Y., et al.: Intelligent energy-based modified super twisting algorithm and factional order PID control for performance improvement of PMSG dedicated to tidal power system. IEEE **9**, 57414–57425 (2021). https://doi.org/10.1109/ACCESS.2021.3072332
18. Karnik, N.N. et al.: Type-s2 fuzzy logic systems. IEEE Trans. Fuzzy Syst. **7**, 643–658 (1999). https://doi.org/10.1109/91.811231
19. De, A.K. et al.: Literature review on type-2 fuzzy set theory. In: Soft Computing, vol. 26, pp. 9049–9068. Springer (2022). https://doi.org/10.1007/s00500-022-07304-4
20. Karnik, N.N., et al.: Type-2 fuzzy logic systems: type-reduction. In: SMC'98 Conference Proceedings. Published on IEEEXplore-August 2002, pp. 2046–2051 (1998). https://doi.org/10.1109/ICSMC.1998.728199
21. Tzounas, G., et al.: Theory and implementation of fractional order controllers for power system applications. IEEE Trans. Power Syst. **35**(6), 4622–4631 (2020). https://doi.org/10.1109/TPWRS.2020.2999415
22. Mittal, K., et al.: A comprehensive review on type-2 fuzzy logic applications: Past, present and future. Elsevier, Eng. Appl. Artif. Intell. **95**(103916), 1–12 (2020)
23. Oldham K.B., Spanier, J.: The Fractional Calculus Theory and Applications of Differentiation and Integration to Arbitrary Order. Elsevier (2006)
24. Miller K. S.,Ross B.,An Introduction to the Fractional Calculus and Fractional Differential Equations. Wiley-Blackwell Publisher (1993)
25. Hanif, O., Kedia, V.: Evolution of Proportional Integral Derivative Controller, ICRIEECE, IEEE Xplore, pp. 2655–2659 (2020). https://doi.org/10.1109/ICRIEECE44171.2018.9008628
26. Varshney, P., Gupta, S.K.: Implementation of fractional fuzzy PID controllers for control of fractional-order systems. In: Conference: ICACCI, IEEE Xplore, pp. 1322–1328 (2014). https://doi.org/10.1109/ICACCI.2014.6968376
27. El-Bardini, M., et al.: Interval type-2 fuzzy PID controller for uncertain nonlinear inverted pendulum system. ISA Trans. **53**(3), 732–743 (2014). https://doi.org/10.1016/j.isatra.2014.02.007
28. Wang, C. et al.: Tuning fractional order proportional integral differentiation controller for fractional order system. In: Proceedings of the 32nd Chinese Control Conference, 2013, pp. 552-555. Electronic ISBN:978-9-8815-6383-5

29. Kumar, A., Kumar, V.: Performance analysis of optimal hybrid novel interval type-2 fractional order fuzzy logic controllers for fractional order systems. Expert Syst. Appl. **93**, 435–455 (2018). https://doi.org/10.1016/j.eswa.2017.10.033

30. Astrom, K.J.,Hagglund, T.: PID Controllers, Theory, Design and Tuning, 2nd edn. Instrument Society of America (1995)

31. Astrom, K.J., Hagglund, T.: The future of PID control. Control Eng. Pract. **9**(11), 1163–1175 (2001). https://doi.org/10.1016/S0967-0661(01)00062-4

32. Jantzen, J.: Foundations of Fuzzy Control. Wiley, Ltd (2007)0-470-02963-3

33. EdetRezaKatebi, E., Katebi, R.: On fractional-order PID controllers. IFAC-PapersOnLine **51**(4), 739–744 (2018). https://doi.org/10.1016/j.ifacol.2018.06.208

34. Birs, I., et al.: Survey of recent advances in fractional order control for time delay system. IEEE Access, vol. 7, 30951–30965. https://doi.org/10.1109/ACCESS.2019.2902567

35. Bennett, S.: Nicholas Minorsky and the automatic steering of ships. IEEE Control Syst. Mag. **4**, 10–15 (1984). https://doi.org/10.1109/MCS.1984.1104827

36. Bennett, S., The past of PID control, IFAC digital control: past, present and future of PlO control. Terrassa. Spain, pp. 1–11 (2000). https://doi.org/10.1016/S1474-6670(17)38214-9

37. Bennett, S.: Development of the PID controller. IEEE Control Syst. Mag. 58–65 (1994). https://doi.org/10.1109/37.248006

38. Ray, P.K., et al.: A hybrid firefly-swarm optimized fractional order interval type-2 fuzzy PID-PSS for transient stability improvement. IEEE Trans. Indust. Appl. **55**(6), 6486–6498 (2019)

39. Sain, D., Praharaj, M., Bosukonda, M.M.: A simple modelling strategy for integer order and fractional order interval type-2 fuzzy PID controllers with their simulation and real-time implementation. Expert Syst. Appl. **202**, 117–196 (2022)

Artificial Bee Colony Optimized Precompensated Interval Type-2 Fuzzy Logic Controller for a Magnetic Levitation System

Anupam Kumar, Ritu Raj, Prashant Gaidhane, and Oscar Castillo

Abstract Magnetic Levitation System (MLS) is highly nonlinear and unstable complex system wherein the load disturbance and external noise adversely affect the system's performance. Therefore, controller design for controlling the ball position is a complex task for researchers. In this chapter, a novel two-layer interval type-2 fuzzy logic controller (TL-IT2FLC) approach is presented for MLS to suspend a steel ball in air without any mechanical support. For optimal controller design, well known optimization technique i.e., artificial bee colony (ABC) is employed to obtain the required control parameters. In addition, the MLS is also tested in presence of disturbance, varying reference positions, and random noise to show the efficacy of the proposed method. Finally, the simulation results clearly show that the performance of TL-IT2FLC approach is superior to a two-layer type-1 fuzzy logic controller (TL-T1FLC), and a conventional T1FLC.

Keywords Interval type-2 fuzzy logic controller · Two-layer fuzzy logic controller · Magnetic levitation system · Artificial bee colony · Robustness testing

A. Kumar (✉)
Department of Electronics and Communication Engineering, National Institute of Technology Patna, Patna, India
e-mail: anuanu1616@gmail.com; anupam.ec@nitp.ac.in

R. Raj
Department of Electronics and Communication Engineering, Indian Institute of Information Technology Kota, Jaipur, India
e-mail: riturajsam@gmail.com

P. Gaidhane
Deparment Instrumentation Engineering Department GCOE, Jalgaon, India
e-mail: pjgaidhane@gmail.com

O. Castillo
Tijuana Institute Technology, Tijuana, Mexico
e-mail: ocastillo@tectijuana.mx

1 Introduction

Magnetic Levitation System (MLS) is a highly nonlinear, unstable, and uncertain system. It is mainly due to the nonlinear open-loop unstable complex dynamics but lower order model can be calculated via linearization. The obtained model easily handled with linear/nonlinear methods. Uncertainties come in the system because of modeling error, electromagnetism interference, and other outside disturbances. Due to these uncertainties, the position control of the levitated object becomes very difficult. Thus, it is essential to design intelligent controllers for better performance and accurate positioning of levitated objects. The concept of MLS can be utilized for various applications such as maglev vehicle, wild tunnel magnetic levitation system, high-speed maglev motor, magnetic levitation anti-vibration system, and maglev trains.

Fuzzy logic theory was developed in 1965 [1] and it becomes immense famous in control applications for several real world challenging task. The fuzzy controller is based on human expertise knowledge. For generating enhanced performance, it is classified into three categories: direct action fuzzy logic controller (FLC) type, hybrid FLC type, and gain scheduling FLC type [2]. Furthermore, many control researchers have proposed the hybrid structures of FLCs for different control applications by producing effective controllers. Kim et al. [3] proposed a two-layered fuzzy logic controller (TL-FLC), a combination of fuzzy pre-compensator followed by conventional FLC for the system having deadzone. It provided better performance than conventional fuzzy controllers. Kumar and Raj [4] presented fractional order TL-FLC for controlling mean arterial blood pressure for blood pressure control and proposed controller perform found better than conventional controllers. Most recently, Sharma and Kumar [5] proposed a two-layered interval type-2 FLC (TL-IT2FLC) design for regulating mean arterial blood pressure using SNP drug delivery. Here, the precompensated IT2FLC was introduced in front of the main IT2FLC controller that reduces the effect of uncertainties and external disturbances. Further, the simulation results claimed that the proposed controller produced better results than existing methods. From the above discussion, it clear that the proposed TL-IT2FLC controller can also be employed to MLS that is highly complex in nature.

For the past few decades, it has been shown in several research works that IT2FLCs have produced better results than its conventional counterparts because of the additional degree of freedom presented in the membership functions (MFs) due to footprint of uncertainty (FOU) [6–8]. Kumar and Kumar [9] studied the interval type-2 fuzzy PID (IT2FPID) control approach for controlling redundant robot manipulators while tuning the control parameters was done using genetic algorithms. From the simulation results, it was observed that the proposed controller was superior to the conventional T1FLC and PID controllers. Kumbasar and Hagras [10] proposed IT2FPID controller for cascade control structure for path tracking problem of mobile robot application via Big Bang–Big Crunch (BB–BC) optimization method. Due to its prominent performance, it is reported in many control applications [11–14].

Considering the literature above, it is apparent that the type-2 fuzzy controller's performance has been superior to its conventional counterparts.

In this chapter, a new TL-IT2FLC controller is demonstrated for a complex MLS which is high prone to uncertainties. Further, the number of control parameters of TL-IT2FLC, TL-T1FLC and single layered T1FLC (SL-T1FLC) are comparatively high and their tuning cannot be easily done by trial-and-error methods, therefore, optimization techniques based on heuristics are used to tune the controllers. In this work we employ ABC optimization technique to tune the controller parameters. Here, the total design parameters of TL-IT2FLC, TL-T1FLC and SL-T1FLC is eight, eight and four respectively. It is a complicated process for tuning such large number of control parameters. Further, it is seen that optimizing all three specifications: MFs, fuzzy rules and scaling factors is a cumbersome process. Therefore, the few control parameters are taken fixed such as scaling factors and rule base for showing the importance of the proposed controllers, whereas, the MFs are upgraded to type-2 MFs. Therefore, the proposed TL-IT2FLC approach is compared to TL-T1FPID and SL-T1FLC for ball tracking and shows better results.

The rest of this chapter is organized as follows: Section 2 provides the mathematical description of MLS. Section 3 presents the details of the proposed TL-IT2FLC scheme and in Sect. 4 the basics of ABC optimization technique is presented. At last, Sects. 5 and 6 describe the simulation results and conclusions, respectively.

2 Mathematical Analysis of Magnetic Levitation System

A electromagnetic suspension system is a platform for performing magnetic suspension experiments where a steel ball is levitating in the air with no mechanical support. The basic block diagram of the MLS is shown in Fig. 1. The working of MLS is given as: when electric current goes through the windings, an electromagnetic force is generated which pulls the suspended object upwards. The current in the coil is adjusted such that it generates a force just sufficient to counterbalance the weight of the object at some equilibrium point. Thus, the object is suspended at the corresponding equilibrium point. For stable control of the object position, feedback loop is used. The distance of the object from the bottom of the electromagnet is measured by a light source and sensor assembly. The signal from the sensor is used to generate a control voltage using a digital controller, which drives a suitable current through the coil.

The mathematical modeling of MLS is described in this section. Here, we have considered that the steel ball is not under the influence of any external forces except electromagnetic force (F) and gravitational force (mg). Along with this, it is also assumed that there is no influence of inductance on current flowing in the electromagnet and is considered as control current. The output current of power amplifier is assumed to be linear and in phase of input voltage. Thus, the dynamic equation in vertical direction can be expressed [7, 11] by the following equations,

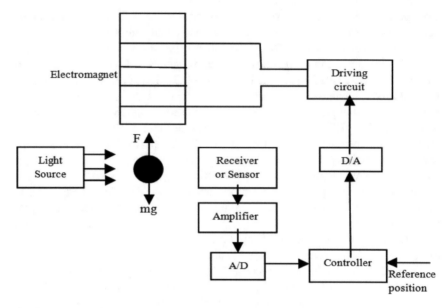

Fig. 1 Basic block diagram of MLS system

$$m\frac{d^2x}{dt^2} = \frac{2Ki_0}{x_0}i - \frac{2Ki_0^2}{x^3}x \tag{1}$$

After applying Laplace transformation to (1), we get

$$x(s)s^2 = \frac{2Ki_0}{mx_0^2}i(s) - \frac{2Ki_0^2}{mx_0^3}x(s) \tag{2}$$

Now, applying boundary condition $mg = -K(i_0/x_0)^2$ then open-loop transfer function can be expressed as

$$\frac{x(s)}{i(s)} = \frac{-1}{As^2 - B} \tag{3}$$

where, x = the position of the steel ball levitating from reference level; m = mass of steel ball, i = current in electromagnetic coil, g = acceleration due to gravity, x_0 = reference position of the steel ball, and A $= i_0/2g$, B $= i_0/x_0$.

The input voltage (V_{in}) of the power amplifier is defined as input variable, and the output voltage (V_{out}) of sensor is referred as output variable. Here, the output voltage provides exact information of the ball position. Thus, mathematical modeling of MLS can be expressed as

$$G(s) = \frac{V_{out}(s)}{V_{in}(s)} = \frac{K_s x(s)}{K_a i(s)} = \frac{-\left(\frac{K_s}{K_a}\right)}{As^2 - B} \qquad (4)$$

where, K_a is power amplifier gain and K_s is sensor amplifier gain.

After getting transfer function of MLS, we can get state-space variables. Then the state variables are defined as $x_1 = V_{out}$ and $x_2 = \dot{V}_{out}$. The system's state-space equation is given as

$$\begin{bmatrix} \dot{x}_1 \\ \dot{x}_2 \end{bmatrix} = \begin{bmatrix} 0 & 1 \\ \frac{2g}{x_0} & 0 \end{bmatrix} \begin{bmatrix} x_1 \\ x_2 \end{bmatrix} + \begin{bmatrix} 0 \\ -\frac{2gK_s}{i_0 K_a} \end{bmatrix} v_{in} \qquad (5)$$

$$y = \begin{bmatrix} 1 & 0 \end{bmatrix} \begin{bmatrix} x_1 \\ x_2 \end{bmatrix} = x_1 \qquad (6)$$

The dynamics of MLS is constructed using (5) and (6) with specifications from [7, 11].

$$\begin{bmatrix} \dot{x}_1 \\ \dot{x}_2 \end{bmatrix} = \begin{bmatrix} 0 & 1 \\ 653.4 & 0 \end{bmatrix} \begin{bmatrix} x_1 \\ x_2 \end{bmatrix} + \begin{bmatrix} 0 \\ 2499.1 \end{bmatrix} u_{in} \qquad (7)$$

$$y = x_1 \qquad (8)$$

A few fundamental meanings of different strategies, fuzzy numbers are given in this part which will be essential in the current study.

3　Two-Layered Interval Type-2 Fuzzy Logic Controller

In this section, detailed design and implementation of the TL-IT2FLC scheme are introduced along with rules and MFs. TL-IT2FLC scheme comprises of an interval type-2 fuzzy pre-compensator controller (IT2FPC) and the conventional interval type-2 fuzzy proportional derivative (IT2FPD) controller. The IT2FPC is inserted to change the control signal, to compensate for overshoot and steady-state error in the output response and meliorate the output performance to counteract the load disturbances and measurement noise. Figure 2 demonstrates the TL-IT2FLC scheme for levitating a steel ball at a given reference position. The presented controller, i.e., TL-IT2FLC is inspired by a recent paper [5]. The mathematical analysis of the TL-IT2FLC is given as:

The control law of IT2FPC is given as:

$$U_p(t) = IT2FLC1\big(e_p(t), \dot{e}_p(t), U_p(t-1)\big) \qquad (9)$$

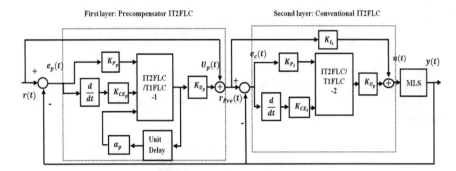

Fig. 2 The structural diagram of TL-IT2FLC scheme for MLS

where $U_p(t)$ is the output of IT2FPC, and the three input is considered as $e_p(t)$, $\dot{e}_p(t)$, and $U_p(t-1)$.

Now, the changed reference position $r_{pre}(t)$ and the new (updated) error $e_t(t)$ for IT2FPD can be given as:

$$r_{pre}(t) = U_p(t) + r(t) \tag{10}$$

$$e_t(t) = r_{pre}(t) - y(t) \tag{11}$$

In the end, the input to the MLS, i.e., final control output is obtained as below:

$$u(t) = IT2FLC2(e_t(t), \dot{e}_t(t)) + K_{I_t} r_{pre}(t) \tag{12}$$

where K_{I_t} is the feed-forward gain. For the first layer, the terms K_{P_p}, K_{CE_p}, K_{U_p} and α_p are the scaling factors for IT2FPC and the terms K_{P_t}, K_{CE_t} and K_{U_t} are the scaling factors for IT2FPD controller.

In the TL-IT2FLC scheme, the two layers of IT2FLCs are utilized. Here, the pre-processing block's purpose is to scale input and output variables into physical range [15, 16]. Further, the TL-IT2FLC scheme consists of two layers and is given as follows [5]:

(1) *First Layer:* In the present study, the two IT2FLCs are utilized for each layer for MLS applications. In the IT2FPC, there are three input variables, namely $e_p(t)$, $\dot{e}_p(t)$, and $U_p(t-1)$ and a single output variable, namely $U_p(t)$. Here, the input MFs of the first layer are given as: Negative (NE_{MLS}), Zero (Z_{MLS}) and Positive (PO_{MLS}) with the specification in the MFs as $m_N=0.2$, $m_Z=0.9$ and $m_P=0.2$ [17]. On the other hand, the output MFs are distinguished by seven singleton such as Negative Big (NB_MLS) $= -1$, Negative Medium (NM_MLS) $= -0.6667$, Negative Small (NS_{MLS}) $= -0.333$, Zero (Z_{MLS}) $= 0$, Positive Medium (PM_{MLS}) $= 0.667$, Positive Small (PS_{MLS}) $= 0.333$, and

Positive Big (PB_{MLS}) = 1. For the IT2FPC, the rule base with three input variables and a single output variable and three triangular type MFs will produce 27 rules [3] as given in Table 1.

(2) *Second Layer:* The second layer of the proposed controller is traditional IT2FPD controller as shown in Fig. 2. The inputs of standard IT2FPD controller are described by three triangular MFs, namely Negative (NE_{MLS}), Zero (Z_{MLS}) and Positive ($P\,O_{MLS}$) as shown in Fig. 3a. Whereas, the output MFs are distinguished by five singletons such as Negative (N_{MLS}) = -1, Negative Medium

Table 1 Rule base for IT2FLC/T1FLC of first layer

IF			THEN
$e_p(t)$	$\dot{e}_p(t)$	$U_p(t-1)$	$U_p(t)$
NE_{MLS}	NE_{MLS}	NE_{MLS}	NS_{MLS}
		Z_{MLS}	Z_{MLS}
		$P\,O_{MLS}$	Z_{MLS}
	Z_{MLS}	NE_{MLS}	PS_{MLS}
		Z_{MLS}	Z_{MLS}
		$P\,O_{MLS}$	NS_{MLS}
	$P\,O_{MLS}$	NE_{MLS}	PM_{MLS}
		Z_{MLS}	PS_{MLS}
		$P\,O_{MLS}$	Z_{MLS}
Z_{MLS}	NE_{MLS}	NE_{MLS}	Z_{MLS}
		Z_{MLS}	NS_{MLS}
		$P\,O_{MLS}$	NS_{MLS}
	Z_{MLS}	NE_{MLS}	Z_{MLS}
		Z_{MLS}	Z_{MLS}
		$P\,O_{MLS}$	Z_{MLS}
	$P\,O_{MLS}$	NE_{MLS}	PS_{MLS}
		Z_{MLS}	PS_{MLS}
		$P\,O_{MLS}$	Z_{MLS}
$P\,O_{MLS}$	NE_{MLS}	NE_{MLS}	PM_{MLS}
		Z_{MLS}	PS_{MLS}
		$P\,O_{MLS}$	Z_{MLS}
	Z_{MLS}	NE_{MLS}	PM_{MLS}
		Z_{MLS}	PS_{MLS}
		$P\,O_{MLS}$	Z_{MLS}
	$P\,O_{MLS}$	NE_{MLS}	PB_{MLS}
		Z_{MLS}	PS_{MLS}
		$P\,O_{MLS}$	Z_{MLS}

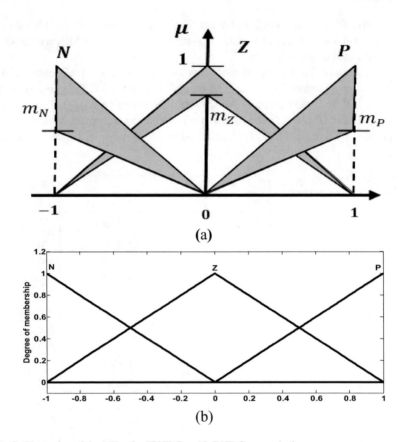

Fig. 3 Illustration of the MFs of **a** IT2FLC and **b** T1FLC respectively

Table 2 Rule base for IT2FLC/T1FLC of second layer

Error/rate of change of error	NE_{MLS}	Z_{MLS}	PO_{MLS}
NE_{MLS}	N_{MLS}	NM_{MLS}	Z_{MLS}
Z_{MLS}	NM_{MLS}	Z_{MLS}	PM_{MLS}
PO_{MLS}	Z_{MLS}	PM_{MLS}	P_{MLS}

$(NM_{MLS}) = -0.8$, Zero $(Z_{MLS}) = 0$, Positive Medium $(PM_{MLS}) = 0.8$, Positive $(P_{MLS}) = 1$. Here, the total number of rules is 9 as given in Table 2 [10, 17].

4   A Brief on Artificial Bee Colony

In this section, the well-known artificial bee colony (ABC) optimization technique proposed by Karaboga [18] is briefly described. Different techniques were proposed in the literature on the concept of honey bees; among them, ABC has become quite popular in the research community [19]. It is applied in many engineering applications such as power system application [20], wireless sensor networks [21], control systems [22], artificial intelligence [23] etc.

ABC optimization technique proposed in [24] has mainly three categories of bees, i.e., employed bees, onlooker bees, and scout bees. For optimization, ABC method requires the following four phases.

4.1 Initialization of Phase

In this phase, the initialization of the population is done randomly in the given search range (U, L) using Eq. (13).

$$x_j^i = L_j^{min} + \text{rand}(0, 1)\left(U_j^{max} - L_j^{min}\right) \tag{13}$$

where, $i = 1,..,N_{food}$ (number of food sources) and $j = 1,..,D_p$ (dimension of the optimization parameter).

4.2 Employee Bee Phase

Each employee bee utilizes a food source for which the bee is selected. It searches for a new candidate solution in its vicinity by Eq. (14).

$$v_j^i = x_j^i + ran[-1, 1](x_j^i - x_j^m) \tag{14}$$

where m and j are mutually exclusive in the proposed techniques.

4.3 Onlooker Bee Phase

The role of the onlooker bee is to choose the food sources based on the dance performed by the employed bees in the hive. Once the food sources are selected probabilistically, which are given as:

$$p_i = 0.1 + \frac{0.9 \times F(X^i)}{max(F(X^i))} \tag{15}$$

where $X^i = (x_1^i, \ldots . x_{D_p}^i)$,

Then, the onlooker bee works similarly to the employed bee acts in its phase.

4.4 Scout Bee Phase

In every cycle, the limits of trial counters of all solutions are examined after both phases i.e., employee bees and onlooker bees. Here, said limit is exceeded the max values, then, abandoned solution can be produced. The linked employed bees became scout. After that scout bee will generate new solution with Eq. (13). In this chapter, the performance index (objective function) for the proposed controller is chosen as an integral absolute error (IAE) for all simulation work, i.e.,

$$F = \int |e(t)| dt \tag{16}$$

5 Simulation Results

In this section, the effectiveness of this new enhanced TL-IT2FLC approach is presented to control MLS. The performance analysis and robustness testing of the presented control approach are done along with other conventional TL-T1FLC and SL-T1FLC controllers. The presented controllers and magnetic levitation system are implemented in the simulink environment of MATLAB as illustrated in Fig. 2. The optimal controller parameters for TL-T1FLC and SL-T1FLC approach are listed in Table 3 using ABC optimization technique. In order to accomplish a reasonable assessment, the rule base are taken same for TL-T1FLC and TL-IT2FLC for step response as well as load disturbance and noise rejection analysis.

In the first performance study, the step response and the load disturbance (applied at t = 1 s) for the MLS is examined and shown that the proposed controller produces better results. The step response along with load disturbance and controller output of the presented controllers are shown in Fig. 4. From the Fig. 4 and Table 4, it is

Table 3 Optimal controllers parameters

Parameters	Controller TL-T1FLC	Parameters	Controller SL-T1FLC
K_{P_p}	84.8017	K_p	8.5098
K_{CE_p}	0.2687	K_{CE}	0.0742
K_{U_p}	0.0338	K_{PD}	0.7010
α_p	0.8494	K_{PI}	29.8920
K_{P_t}	9.9465	–	–
K_{CE_t}	0.2320	–	–
K_{U_t}	0.9936	–	–
K_{I_t}	0.0059	–	–

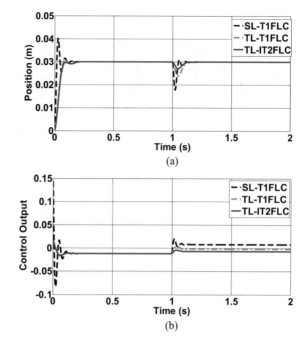

Fig. 4 Performance comparison of the **a** Output response and **b** control signal for SL-T1FLC, TL-T1FLC, and TL-IT2FLC with load disturbance

Table 4 Summary of performance comparison of controllers

Controller	IAE	
	Step + Load disturbance	Random Noise + Load disturbance
TL-IT2FLC	0.01065	0.01226
TL-T1FLC	0.02274	0.02532
SL-T1FLC	0.07496	0.07898

evident that TL-IT2FLC structure produces better results in terms of lesser over-shoots, smaller IAE values, and best load disturbance suppression. For instance, the varying reference position at different levels is also considered as illustrated in Fig. 5.

Moreover, the robustness against noise rejection and external load disturbance rejection of the enhanced TL-IT2FLC scheme is also investigated. At first, the effect of adding random noise of maximum amplitude ± 0.01 (as illustrated in Fig. 6c) at the system's output is studied. Further, the simulation is again done for the varying reference position with noise. The output responses with the controllers are presented in Fig. 6. Therefore, it is evident from the figure that TL-IT2FLC is more robust against random noise compared to TL-T1FLC and SL-T1FLC schemes.

Fig. 5 Performance
comparison of the **a** Output
response **b** control signal of
varying reference position
for SL-T1FLC, TL-T1FLC,
and TL-IT2FLC

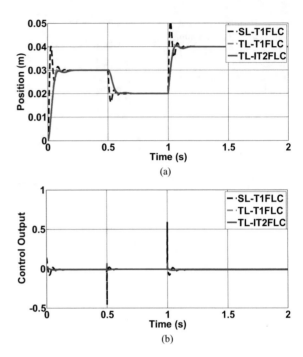

6   Conclusions

In this chapter, a TL-IT2FLC scheme is proposed to control MLS for accurate positioning of a steel ball in air. The proposed controller is a combination of IT2FPC and
conventional IT2FLC. The IT2FPC is utilized to overcome the problem of change in
control signal due to the effect of external disturbance, random noise such as background light, unwanted electromagnetic induction as well as disturbances generated
by processing circuit. An interval type-2 fuzzy logic controller has been incorporated to enhance the closed-loop response of the proposed controller. The comparative study has been executed to depict that the TL-IT2FLC approach gives better
response than TL-T1FPID and SL-T1FLC for step response with load disturbance
and varying reference positions. At the end of simulation, the robustness analysis is
also done for load disturbance and noise.

Fig. 6 Illustration of the **a** Output response of step input along with load disturbance **b** Output response of varying reference position **c** random noise for SL-T1FLC, TL-T1FLC, and TL-IT2FLC for the addition of noise in the feedback path

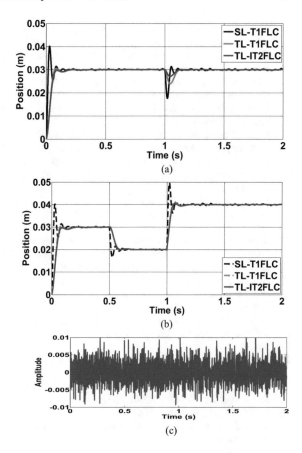

Acknowledgements Authors declared that there is no financial support for this work.

References

1. Zadeh, L.A.: The concept of a linguistic variable and its application to approximate reasoning-I. Inf. Sci. (Ny) **8**, 199–249 (1975). https://doi.org/10.1016/0020-0255(75)90036-5
2. Hu, B.-G., Mann, G.K.I., Gosine, R.G.: A systematic study of fuzzy PID controllers-function-based evaluation approach. IEEE Trans. Fuzzy Syst. **9**, 699–712 (2001). https://doi.org/10.1109/91.963756
3. Kim, J.H., Park, J.H., Lee, S.W., Chong, E.K.: A two-layered fuzzy logic controller for systems with deadzones. IEEE Trans. Ind. Electron. **41**, 155–162 (1994). https://doi.org/10.1109/41.293875
4. Kumar, A., Raj, R.: Design of a fractional order two layer fuzzy logic controller for drug delivery to regulate blood pressure. Biomed. Signal Process. Control. **78** (2022)

5. Sharma, R., Kumar, A.: Optimal Interval type-2 fuzzy logic control based closed-loop regulation of mean arterial blood pressure using the controlled drug administration. IEEE Sens. J.
   22, 7195–7207 (2022). https://doi.org/10.1109/JSEN.2022.3151831
6. Hagras, H.: A hierarchical type-2 fuzzy logic control architecture for autonomous mobile
   robots. IEEE Trans. Fuzzy Syst. **12**, 524–539 (2004)
7. Kumar, A., Panda, M.K., Kundu, S., Kumar, V.: Designing of an interval type-2 fuzzy logic
   controller for magnetic levitation system with reduced rule base. In: Computer Communication
   Network Technology, India, pp. 1–8 (2012)
8. Raj, R., Mohan, B.M., Lee, D.E., Yang, J.M.: Derivation and structural analysis of a three-input
   interval type-2 TS fuzzy PID controller. Soft Comput., 1–15 (2022)
9. Kumar, A., Kumar, V.: Evolving an interval type-2 fuzzy PID controller for the redundant
   robotic manipulator. Expert Syst. Appl. **73**, 161–177 (2017). https://doi.org/10.1016/j.eswa.
   2016.12.029
10. Kumbasar, T., Hagras, H.: Big Bang-Big Crunch optimization based interval type-2 fuzzy PID
    cascade controller design strategy. Inf. Sci. (Ny) **282**, 277–295 (2014). https://doi.org/10.1016/
    j.ins.2014.06.005
11. Kumar, A., Panda, M.K., Kumar, V.: Design and implementation of interval type-2 single input
    fuzzy logic controller for magnetic levitation system. In: International Conference Advance
    Computer (ICAdC), ASIC 174, pp. 833–840 (2013)
12. Kumar, A., Kumar, V.: Performance analysis of interval type-2 FSM controller applied
    to a magnetic levitation system. In: International Conference Soft Computer Technology
    Implementations, India, pp. 107–112 (2015)
13. Castillo, O., Melin, P.: A review on the design and optimization of interval type-2 fuzzy
    controllers. Appl. Soft Comput. J. **12**, 1267–1278 (2012). https://doi.org/10.1016/j.asoc.2011.
    12.010
14. Raj, R., Mohan, B.M.: General structure of Interval Type-2 fuzzy PI/PD controller of Takagi-
    Sugeno type. Eng. Appl. Artif. Intell. **87**, 103273 (2020)
15. Mendel, J.M., John, R.I., Liu, F.: Interval type-2 fuzzy logic systems made simple. IEEE Trans.
    Fuzzy Syst. **14**, 808–821 (2006). https://doi.org/10.1109/TFUZZ.2006.879986
16. El-Bardini, M., El-Nagar, A.M.: Interval type-2 fuzzy PID controller for uncertain nonlinear
    inverted pendulum system. ISA Trans. **53**, 732–743 (2014). https://doi.org/10.1016/j.isatra.
    2014.02.007
17. Kumbasar, T., Hagras, H.: Interval type-2 fuzzy PID controllers. In: Janusz, K., Pedrycz, W.
    (Eds.), Springer Handb. Computer Intelligent, pp. 285–294. Berlin, Heidelberg (2015)
18. Karaboga, D.: An idea based on Honey Bee Swarm for numerical optimization, Tech. Rep.
    TR06, Erciyes Univ. (2005) 10.citeulike-article-id:6592152
19. Karaboga, D., Gorkemli, B., Ozturk, C., Karaboga, N.: A comprehensive survey: artificial bee
    colony (ABC) algorithm and applications. Artif. Intell. Rev. **42**, 21–57 (2014). https://doi.org/
    10.1007/s10462-012-9328-0
20. Gozde, H., Taplamacioglu, M.C.: Comparative performance analysis of artificial bee colony
    algorithm for automatic voltage regulator (AVR) system. J. Franklin Inst. **348**, 1927–1946
    (2011). https://doi.org/10.1016/j.jfranklin.2011.05.012
21. Udgata, S.K., Sabat, S.L., Mini, S.: Sensor deployment in irregular terrain using artificial bee
    colony algorithm. Nat. Biol. Inspired Comput. 2009. NaBIC 2009. World Congr., 1309–1314
    (2009). https://doi.org/10.1109/NABIC.2009.5393734
22. Rajasekhar, A., Jatoth, R.K., Abraham, A.: Design of intelligent PID/PI??D?? speed controller
    for chopper fed DC motor drive using opposition based artificial bee colony algorithm. Eng.
    Appl. Artif. Intell. **29**, 13–32 (2014). https://doi.org/10.1016/j.engappai.2013.12.009
23. Huang, H.-C., Chuang, C.-C.: Artificial bee colony optimization algorithm incorporated with
    fuzzy theory for real-time machine learning control of articulated robotic manipulators. IEEE
    Access. **8**, 192481–192492 (2020)
24. Karaboga, D., Basturk, B.: On the performance of artificial bee colony (ABC) algorithm. Appl.
    Soft Comput. J. **8**, 687–697 (2008). https://doi.org/10.1016/j.asoc.2007.05.007

Application of Type-2 Fuzzy Logic Controllers in Renewable Energy Systems

Suraparaju Krishnama Raju

Abstract Handling the uncertainties in the input and controller parameter tuning is a major challenge in the grid integrated operation of renewable energy systems. This chapter proposes a novel control strategy based on Type-2 fuzzy logic sets (T2-FLSs) which can handle the parameter uncertainties in the inputs as well as the grid disturbances. The unique features in the type-2 fuzzy sets such as foot print of uncertainty and third dimension in the membership function (MF) enables them to model the uncertainties very effectively. A novel control strategy is designed for the grid integrated operation of wind energy system that can tolerate the variations in the wind speed as well as grid side disturbances. The performance of the proposed strategy is validated with the simulation results and also compared with the other conventional controllers.

1 Introduction

The steady rise in energy demand poses a challenge to many nations due to the depletion of fossil fuels and carbon emissions. Renewable energy sources (RES) play a key role in mitigating these concerns and many countries are putting efforts to develop the alternate energy sources especially from wind, solar, bio-mass, tidal etc. [1]. Among all the renewable energy sources, in the recent past wind and solar have drawn a special attention for the reason that bulk amount of energy production is possible [2]. In general, the power distribution from source to load takes place by connecting all the sources and loads to a common network known as utility grid. To export power to the grid, all the connected sources need to be operated in synchronization with the grid parameters like voltage, frequency, and phase etc. All the connected sources are supposed to be in compliance with certain grid standards known as the grid codes.

S. Krishnama Raju (✉)
Visvesvaraya National Institute of Technology Nagpur, Nagpur 440010, India
e-mail: krs@eee.vnit.ac.in; skrajuvnit@gmail.com

© The Author(s), under exclusive license to Springer Nature Switzerland AG 2023
O. Castillo and A. Kumar (eds.), *Recent Trends on Type-2 Fuzzy Logic Systems: Theory, Methodology and Applications*, Studies in Fuzziness and Soft Computing 425,
https://doi.org/10.1007/978-3-031-26332-3_5

The major issue with RES based power plants is, the inputs to these plants such as wind, solar etc., are highly uncertain and is a limitation to export bulk amount of power to the utility grid. In order to adhere the grid codes, the output power from the RES is transformed into different forms with the help of power electronic converters (PEC). Further, in case of wind energy, different topologies are proposed with variety of machines and turbines, to tackle with the uncertainty in the wind speed. However, the control of the PECs stands out as the major issue in the grid integrated operation of any RES. Moreover, the system models with RES, are highly non-linear with parameter uncertainty, and designing the control schemes for such systems needs a comprehensive analysis.

To design an effective control scheme, it is necessary to understand the RES structure and control objectives. In this chapter the application of type-2 fuzzy sets and the design of controller parameters will be discussed in detail. To be more specific, let us consider a doubly fed induction generator (DFIG) based wind power plant which is connected to the utility grid through a two stage power electronic converters, generally known as back to back converters. The basic schematic diagram is shown in Fig. 1.

The turbine is directly coupled to the rotor of the DFIG, and the rotor windings are electrically connected to the PEC known as rotor side converter (RSC). The input to the RSC is supplied through a DC link, which is continuously charged from another PEC known as grid side converter (GSC). The stator of DFIG and GSC are directly connected to the grid.

In this topology, under normal operating conditions i.e. if there is a steady wind speed, then the power flow will be from stator to the grid. In case of any variations in the wind speed, the RSC injects the controlling power to the rotor circuit, so that the stator could continue supplying the power to the grid, without violating the standard grid codes. Further, in case of any faults on the grid side, the DFIG is expected to remain connected to the grid, generally known as fault ride through (FRT)

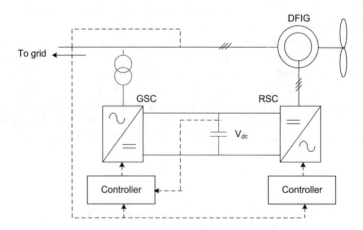

Fig. 1 The single line diagram of the proposed grid connected DFIG

capability. In practice, with reference to DFIG topology although there are multiple control objectives that are to be met, but in this discussion for easy understanding, the control objectives are framed as follows.

- Supplying the power to the grid in variable wind speed conditions and
- Achieving fault ride through capability.

To achieve the above control objectives, the controlling action is initiated through RSC and GSC. Basically in both these converters, the switches are operated at a particular duty ratio to track the output parameters at the reference levels. The control scheme has to decide the appropriate duty ratio tracking the reference parameters which are mostly variable in nature. In case of DFIG topology, there are many subsystems that are interconnected, so the controlling action at any subsystem is very much dependent on the parameters of the other subsystems present in the loop. This makes the system model very complex and requires decentralized control schemes for each subsystem.

2 Control Scheme

The single line diagram of the system with the proposed controllers is shown in Fig. 1. The controller designated for the RSC tracks the reference powers based on the operating conditions. Further, the input to the RSC is the DC source that goes through a change in its voltage levels as and when the RSC is controlled. The controller designated for the GSC has to ensure that the DC voltage levels are maintained at required levels to support the RSC.

In order to understand the system behavior it is necessary to derive the dynamic modelling of the system. For this purpose, the equivalent circuit of the DFIG in an arbitrary reference frame [3] is derived as shown in Fig. 2, where.

v_s—represents the stator voltage
v_r—represents the rotor voltage
R_s—represents the stator resistance
R_r—represents the rotor resistance

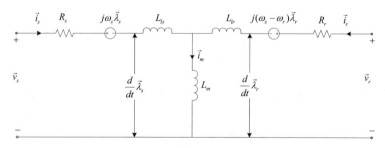

Fig. 2 Equivalent Circuit for modelling the DFIG

ω_r—represents the rotor speed
ω_s—represents the synchronous angular speed
L_{ls}—represents the stator leakage inductances
L_{lr}—represents the rotor leakage inductance
L_m—represents the mutual inductance
λ_s—represents the stator flux
λ_r—represents the rotor flux

The stator and rotor fluxes can be expressed as

$$\lambda_s = L_s i_s + L_m i_r \tag{1}$$

$$\lambda_r = L_r i_r + L_m i_s \tag{2}$$

where,
$L_s = L_{ls} + L_m$ and
$L_r = L_{lr} + L_m$

The voltage across the stator and rotor windings can be derived as

$$v_s = R_s i_s + \frac{d\lambda_s}{dt} + j\omega_s \lambda_s \tag{3}$$

$$v_r = R_r i_r + \frac{d\lambda_r}{dt} + j(\omega_s - \omega_r)\lambda_r \tag{4}$$

In general the synchronous reference frame model is widely used for control analysis of machines, therefore it will be convenient to implement the decoupled control if the flux and voltages are expressed in $d - q$-axis components. The stator and rotor flux are expressed as

$$\lambda_{sd} = L_s i_{sd} + L_m i_{rd} \tag{5}$$

$$\lambda_{sq} = L_s i_{sq} + L_m i_{rq} \tag{6}$$

$$\lambda_{rd} = L_r i_{rd} + L_m i_{sd} \tag{7}$$

$$\lambda_{rq} = L_r i_{rq} + L_m i_{sq} \tag{8}$$

Similarly, the d and q-axis components of the voltages across stator and rotor are

$$v_{sd} = R_s i_{sd} + \frac{d\lambda_{sd}}{dt} - \omega_s \lambda_{sd} \tag{9}$$

$$v_{sq} = R_s i_{sd} + \frac{d\lambda_{sq}}{dt} + \omega_s \lambda_{sd} \tag{10}$$

$$v_{rd} = R_r i_{rd} + \frac{d\lambda_{rd}}{dt} - (\omega_s - \omega_r)\lambda_{rd} \tag{11}$$

$$v_{rq} = R_r i_{rq} + \frac{d\lambda_{rq}}{dt} + (\omega_s - \omega_r)\lambda_{rq} \tag{12}$$

The input power is expressed in terms of active and reactive components at the stator circuit and is derived as

$$P = 1.5(v_{sd}i_{sd} + v_{sq}i_{sq}) \tag{13}$$

$$Q = 1.5(v_{sq}i_{sd} - v_{sd}i_{sq}) \tag{14}$$

The main control objective is to accurately track the given reference powers, even in the presence of various disturbances such as wind speed variation and fault and load changes in the utility grid. To achieve this, the active and reactive powers of the DFIG are controlled, with the help of both the RSC and GSC. Traditionally the control schemes are proposed based on the PI or PID controllers [4], and the limitations of these schemes are explained in the next section.

3 PI/PID Controller

The control scheme to track the output powers of the DFIG with PI controller is shown in Fig. 3, and the rotor currents are considered as controlling variables. The control objectives can be achieved through the RSC, by generating the appropriate firing pulses to its switches. The most popular approach to control the converters is $d - q$ vector control, which uses the cascaded feedback loops to track the given references. The control strategy with stator voltage oriented scheme using PI controllers is considered in this discussion. The RSC generally uses a two stage controller consisting of active power controller and reactive power controller. At first, the voltages and currents that are in 3-phase *abc* form also known as stationary reference frame are converted into synchronous reference frame or $d - q$ form. A three phase, Phase lock loop (PLL) is used for generating the voltage phase angle, which is required for transformation from *abc* to $d - q$ and vice-versa. For a given wind speed the MPPT algorithm decides the appropriate power to be extracted, hence the control scheme has to track this reference power to ensure maximum power from the wind turbine. This is achieved by generating a *d*-axis reference current (i_d) to the PI controller. Similarly the *q*-axis reference current is generated by comparing the reactive power reference and the actual reactive power that is being fed to the grid. The reference $d - q$ axis currents are compared with the measured rotor currents, and the resulted errors are fed as input to the current controllers.

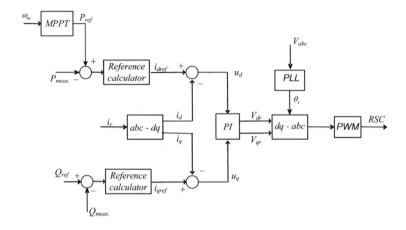

Fig. 3 Control scheme for RSC using PI/PID controller

The controller outputs v_{dr} and v_{qr} are the rotor voltage references in synchronous frame, which are transformed into stationary frame using *dq* to *abc* transformation. The rotor voltages can serve as the three phase reference waveforms to the Pulse width modulator, which generates the gate control pulses to IGBTs that are part of the rotor side converter.

The grid side converter is controlled to maintain the DC link voltage at a constant level and also to control the reactive power when required. The control system of the GSC is shown in Fig. 4. The three phase *abc* components are converted to *dq* components so that the reference signals can be generated in synchronous with the stator voltage. The *dc* link voltage level is maintained with reference to i_{dc} and the reactive power is controlled through the i_{qc} reference. Then, the error signal becomes a reference to the PI/PID controller, whose outputs act as a reference signals for the PWM pulse generator. By comparing the reference signal with a carrier signal the pulse width of the switching pulses can be determined. In general, the PI/PID controller parameters are designed using linearized models, which limits the performance under complex non-linear environment [5, 6]. Moreover, the controller parameters need to be tuned to cope with the variable operating conditions such as faults, wind speed variations etc. [7, 8].

4 Type-1 Fuzzy Logic Controller (Type-1 FLC)

In order to overcome the issues in conventional controller, in [9], a control algorithm based on type-1 fuzzy logic sets (FLSs) is developed to determine the controller parameters. To appreciate the type-1 fuzzy sets, a control scheme with type-1 fuzzy logic controller is designed as shown in Fig. 5. The fuzzy rules are framed based on

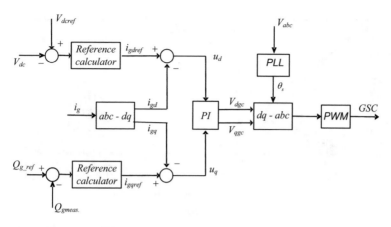

Fig. 4 Control scheme for GSC using PI/PID controller

the quantitative knowledge and experience and the designed membership functions are shown in Fig. 6. The performance of the type-1 FLC is compared with that of conventional PI/PID controllers. In case of typ-1 FLC, the membership functions are not adaptive to minimize the output error for change in operating conditions. Hence always there is trade-off between the performance and the robustness. Further there is an uncertainty in defining the rules, as well as in the membership functions because, once the MFs are chosen for a particular operating conditions, there is a possibility that the degree of MF may vary in accordance with change in operating conditions, which is highly uncertain. Such uncertainties cannot be modelled using type-1 fuzzy sets [10]. Both conventional controllers and type-1 fuzzy controller are unable to handle the parameter uncertainty, which motivates to look for an alternative that can deal with the uncertainty in the controller parameters along with the plant parameter uncertainty.

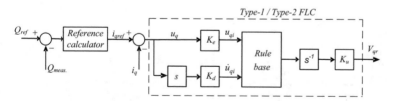

Fig. 5 Control scheme with Type-1/Type-2 FLC

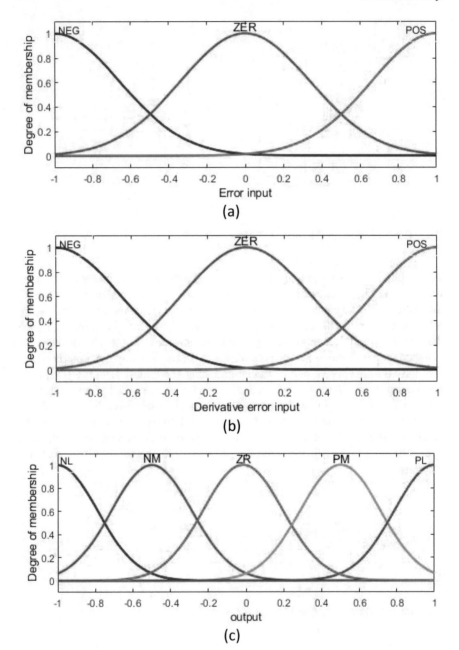

Fig. 6 MFs for type-1 FLC **a** Input1 **b** Input2 **c** Output

5 Proposed Control Scheme with Type-2 FLC

Type-2 fuzzy logic sets, an extension of type-1 fuzzy logic sets, are introduced as a promising tool with a features such as three dimensional membership function and foot print of uncertainty (FOU). The membership grade in any type-1 FLS is defined as a crisp number in [0,1], where as in type-2 FLS, the membership grade for each element in the set is defined as another fuzzy set in [0, 1] [11]. The unique features of type-2 fuzzy sets i.e. the third dimension in the MF and the FOU offers an additional degree of freedom in modeling the uncertainties. The existing control schemes in renewable energy systems are facing serious challenges to upkeep the good performance over wide operating conditions. Hence an attempt has been made to address the limitations of type-1 FLSs, by exploring the type-2 FLSs.

A Type-2 fuzzy set, A, can be characterized by a type-2 fuzzy membership function $\mu_A(x, u)$, where $x \in X$ and $\mu \in J_x \subseteq [0, 1]$ i.e.

$$A = \{(x, u), \mu_A(x, u)\} \ \forall x \in X, \forall u \in J_x \subseteq [0, 1] \tag{15}$$

in which $0 \leq \mu_A(x, u) \leq 1$. The general type-2 FLSs are computationally intensive, hence in this design the interval type-2 fuzzy logic sets (IT-2 FLSs) are used, whose secondary membership grade is considered as unity. An interval type-2 set $\tilde{A}$ is characterized as

$$\tilde{A} = \int_{x \in X} \int_{u \in J_x} 1/(x, u), J_x \subseteq [0, 1] \tag{16}$$

$$= \int_{x \in X} \left[\int_{u \in J_x} 1/u \right] / x \tag{17}$$

where $\tilde{A} : X \rightarrow \{[a, b] : 0 \leq a \leq b \leq 1\}$. The union of all the primary memberships represent the uncertainty of $\tilde{A}$, and can be called as foot print of uncertainty of $\tilde{A}$.

6 Type-2 FLC Design

The proposed control structure with type-2 FLC is shown in Fig. 5. The first step in the design process is to normalize the rotor currents in terms of per unit values and then transform the three phase quantities from *abc* to *dq* components. To calculate the d-q components, the reference frame is derived by taking the angular difference between stator flux vector and physical rotor direct axis. The main advantage of the vector control is the *d* and *q* components of rotor currents would enable independent control of the rotor excitation and electrical torque. At a given wind speed, the amount of active power to be injected is decided by the maximum power tracking algorithm which is taken as the reference value for v_{dr}. The reference value for the v_{qr} is generated from the set point value of the amount of reactive power to be injected. The step by step design procedure for the current controller is explained as follows [12]. The errors to the current controller are defined as

$$u_d = i_{dref} - i_{dr} \tag{18}$$

$$u_q = i_{qref} - i_{qr} \tag{19}$$

and the derivative of the each error is calculated over a fixed time period and are represented as $\dot{u}_d$ and $\dot{u}_q$. Finally, the controller parameters represented by K_e and K_d are chosen to scale the errors in the desired range; similarly the gain K_u is chosen to scale the output without pushing the controller to saturation mode. The final controller inputs after the scaling are expressed as

$$u_{qi} = K_e u_q \tag{20}$$

$$\dot{u}_{qi} = K_d \dot{u}_q \tag{21}$$

In the proposed type-2 FLC design, at first the crisp inputs to the controller are transformed to fuzzy variables using different membership functions as shown in Fig. 7. Based on the magnitude of the error input, each input membership function is mapped to a output membership function and the mapping criterion is defined with set of rules, this process is known as inference mechanism [13]. After the mapping process the derived output is in type-2 form which cannot be used as it is, for the reason that computational burden would be high. To improve the controller response time, the type-2 fuzzy sets are transformed back to type-1 by a process known as type reduction operation. Finally, the type reduced sets are used to calculate the crisp value with the help of different defuzzification techniques, by this process the computational burden or process delay of the controller can be drastically reduced.

The error inputs are fuzzified using three Gaussian membership functions and the output with five MFs. The fuzzy sets for the inputs are defined as Negative-NEG, Zero-ZER, and Positive-POS and for the output as: Negative Large—NL, Negative Medium—NM, Zero—ZR, Positive Large—PL, Positive Medium—PM. The normalized values of the errors and output are defined in the range of $+1$ to -1.The rules for mapping the input and output membership functions are defined using IF–THEN logic as follows.

If u_d is x_1 and $\dot{u}_d$ is y_1 then V_d is w_1

If u_q is x_2 and $\dot{u}_q$ is y_2 then V_q is w_2

By following the above logic, there are 9 rules defined for the entire range of the error and derivative of the error, which are listed in Table 1. In the mapping process of type-2 fuzzy sets, various operations such as meet, join, join aggregation

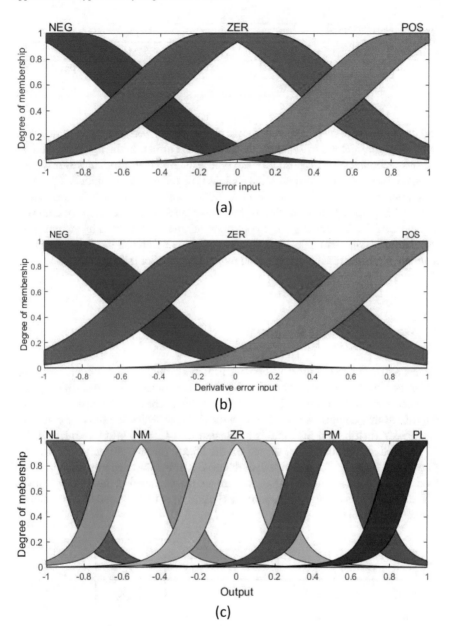

Fig. 7 Membership functions for type-2 FLC **a** Input1 **b** Input2 **c** Output

Table 1 Rules for type-2 FLC

E/ΔE	NEG	ZER	POS
NEG	NL	NM	ZE
ZER	NM	ZE	PM
POS	ZE	PM	PL

and meet implication are involved. In this algorithm, for performing all these operations only min-method and max method are used. Further, in the literature there are various techniques are proposed for the type reduction operation, however the best accuracy has been achieved with the 'height type reduction' technique. To convert the output from fuzzy to crisp value, centroid type defuzzification technique is used. The challenging task in the design is, to choose the appropriate values of controller parameters (K_e, K_d, and K_u). The controller parameters are tuned by analyzing the simulation results, and also by following the rules described in [14]. Further any heuristic algorithm based optimization may be applied for better accuracy.

7 Simulations and Results

To test and validate the proposed type-2 FLC performance, simulations have been performed in a test system as shown in Fig. 8 and a comparative analysis of the results has been done with the type-1 FLC and PI controller. The model consists of a 9 MW wind farm that comprises six 1.5 MW wind turbines which are directly connected to a 25-kV distribution line, and is connected to a 120-kV grid through a 30-km, 25-kV feeder. The load is connected on the 25 kV feeders through a transformer. In order to simulate the network disturbances, a three phase circuit fault on 25-kV feeder and a voltage sag on 120 kV system are programmed in the simulink model. The performances of the system with the type-2, type-1 FLC's and PI controller are plotted and discussed.

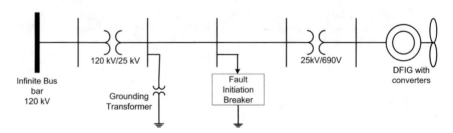

Fig. 8 Grid connected DFIG test system

All the simulations in this work are performed with the following operating conditions; a constant wind speed of 12 m/s, reference terminal voltage and rotor speed at 1 p.u.

8   Fault on 25 kV System

Considering the system in Fig. 8, a three phase fault is initiated for a time period of 0.08 s, and the responses of the system for PI and FLCs is recorded. The performance of all the three controllers is compared in terms of different indexes and listed in Table 2. The responses of the system especially the voltages, currents and powers with PI controller in the loop are shown in Fig. 9. From the results, it is noticed that after initiation of the fault, the voltage level at the stator windings has reduced to 0.1 p.u., soon after the fault clearance it has restored to normal value with damped oscillations. From the power system stability perspective, damping the current oscillations within the prescribed time limits plays an important role in maintaining the grid stability. In this study, both the stator and rotor currents settle after 0.09 s and 0.1 s respectively. The oscillations in the current in turn lead to sustained oscillations in the voltage level of the DC link, and also in the active and reactive powers.

The response of the system with the type-1 FLC is shown in Fig. 10. At the time of initiation of fault the voltage level at the stator windings drops to 0.15 p.u., and during the fault period the stator current rises to 5 times of it rated value, however there is a reduction in the current oscillations compared to that of PI controller. The high currents in the stator windings, caused a fivefold increase in the rotor windings, which is due to strong magnetic coupling between the rotor and stator. The sudden change in the voltage levels further lead to oscillations in the GSC output powers. Further, during the fault period, due to the difference in the input and output powers of the RSC, there is a sharp rise in the voltage level of the dc link to 1.341 p.u., however, the presence of the protection circuit doesn't allow it to deviate further and brought back to normal value. There is a rise in active power amplitude with oscillations peaking to 3.9 p.u. also the reactive power levels with negative peak of 6.1 p.u.

The performance of the system with the type-2 FLC is shown in Fig. 11. During the fault period there is a reduction in the magnitude of the stator currents by 0.85%

Table 2 Performance of PI, Type-1 FLC and Type-2 FLC

Controller	Voltage drop during the fault (p.u)	DC link voltage (p.u.)	Settling time of stator current (s)	Settling time of rotor current (s)
PI	0.1	1.342	0.66	0.76
Type-1 FLC	0.13	1.341	0.36	0.46
Type-2 FLC	0.15	1.34	0.18	0.26

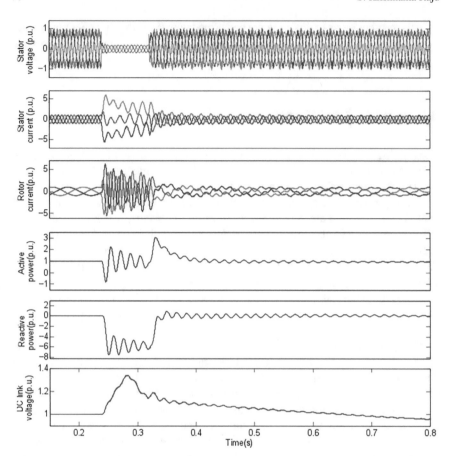

Fig. 9 Responses with PI controller

with an improved settling time. In the PI case, after clearing the fault, the rotor current waveform gets distorted very severely, in contrast with the proposed type-2 FLC it settles within 0.2 s. Also an improvement is observed in the oscillation of the dc link voltage. It is observed that the dynamic response of the controller is fast enough to cope with the standard grid codes, which was seen in terms of non-triggering of the protection system. Further, when compared with PI and type-1 FLC, the proposed type-2 FLC damped out the power oscillations within 0.2 s, which helped the voltage recovery to acceptable limits.

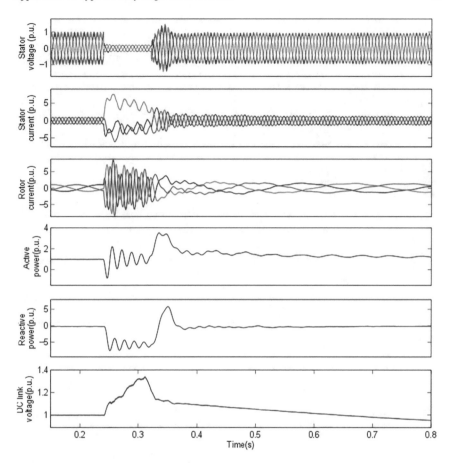

Fig. 10 System response with Type-1 FLC

9 Remote Fault on 120 kV System

In order to analyze the impact of remote fault, a voltage drop of 0.15 p.u. lasting 0.5 s is initiated at t = 7 s on the 120 kV bus. The responses of terminal voltage, rotor current and rotor speed are plotted and discussed. As depicted in Fig. 12, at the instant of fault, with type-1 FLC, the terminal voltage level dropped to 0.85 p.u; after the clearance, the oscillations in the rotor current and speed are settled after 2 s. With the type-2 FLC, the fault voltage is maintained at 0.86 p.u. and quickly recovered to the steady state. The oscillations in the rotor speed and current are damped out much faster to type-1 FLC without degrading the transient performance. This shows that a properly optimized type-2 FLC may perform much better than that of type-1 FLC or PI/PID controllers.

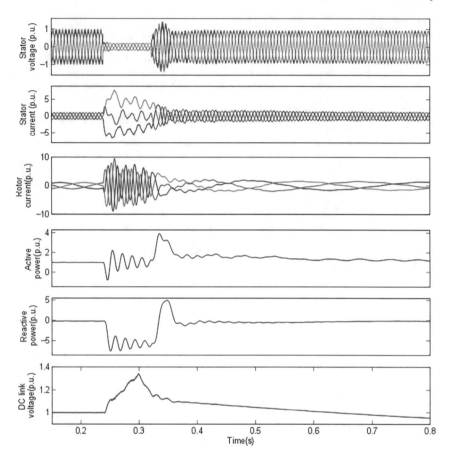

Fig. 11 Responses with Type-2 FLC

10 Conclusion

In this chapter, a novel Type-2 fuzzy sets based control scheme is proposed for the grid integrated operation of DFIG. The design process of the proposed controller is explained and is tested for various disturbances such as three phase short circuit fault, wind speed variations etc. Further, a comparative analysis is done with the PI and Type-1 FLC. The analysis of the simulation results showed that the dynamic response of the type-2 FLC is fast enough to cope with the standard grid codes. Moreover, the proposed type-2 FLC based control strategy is able to effectively tolerate the network faults and input variations, adapting to the change in operating conditions. The analysis in this chapter proves that the type-2 FLC is a valid option for controlling the systems that have uncertainty in the parameters.

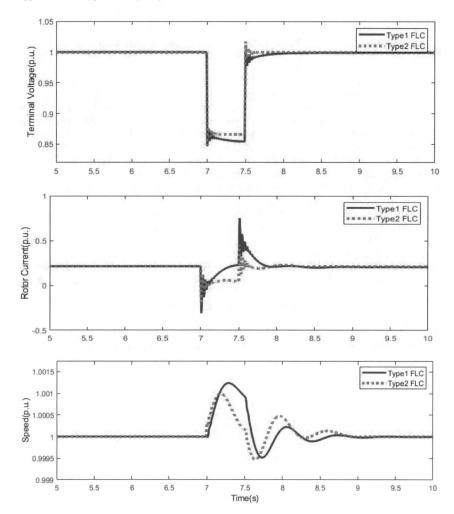

Fig. 12 Responses for a remote fault

References

1. Xie, D., Xu, Z., Yang, L., Ostergaard, J., Xue, Y., Wong, K.P.: A comprehensive LVRT control strategy for DFIG wind turbines with enhanced reactive power support. IEEE Trans. Power Syst. **28**(3), 3302e3310 (2013)
2. Kamel, R.M., Chaouachi, A., Nagasaka, K.: Wind power smoothing using fuzzy logic pitch controller and energy capacitor system for improvement microgrid performance in islanding mode. Energy **35**(5), 2119e2129 (2010)
3. Pena, R., Clare, J., Asher, G.: Doubly fed induction generator using back-to-back PWM converters and its application to variable-speed wind-energy generation. IEEE Proc. Electr. Power Appl. **143**(3), 231 (1996)

4. Krishnama Raju, S., Pillai, G.N.: Design and real time implementation of type-2 fuzzy vector control for DFIG based wind generators. Renew. Energy **88**, 40–50 (2016)
5. Hansen, A.D., Michalke, G.: Fault ride-through capability of DFIG wind turbines. Renew. Energy **32**(9), 1594e1610 (2007)
6. Li, S., Haskew, T.A.: Characteristic study of vector-controlled doubly-fed induction generator in stator-flux-oriented frame. Electr. Power Comp. Syst. **36**(9), 990e1015 (2008)
7. Jacomini, R.V., Franca, A.P., Bim, E.: Simulation and experimental studies on double-fed induction generator power control at sub synchronous operating speed. In: 2009 International Conference on Power Electronics and Drive Systems (PEDS), IEEE, pp. 1421e1424 (2009)
8. Yao, J., Li, H., Chen, Z., Xia, X., Chen, X., Li, Q., Liao, Y.: Enhanced control of a DFIG based wind-power generation system with series grid-side converter under unbalanced grid voltage conditions. IEEE Trans. Power Electron. **28**(7), 3167e3181 (2013)
9. Jabr, H.M., Lu, D., Kar, N.C.: Design and implementation of neuro-fuzzy vector control for wind-driven doubly-fed induction generator. IEEE Trans. Sustain. Energy **2**(4), 404–413 (2011)
10. Hagras, H., Wagner, C.: Towards the wide spread use of type-2 fuzzy logic systems in real world applications. IEEE Comput. Intell. Mag. **7**(3), 14–24 (2012)
11. Mendel, J.M., John, R.I., Liu, F.: Interval type-2 fuzzy logic systems made simple. IEEE Trans. Fuzzy Syst. **14**(6), 808–821 (2006)
12. Krishnama Raju, S., Pillai, G.N.: Design and implementation of Type-2 fuzzy logic controller for DFIG-based wind energy systems in distribution networks. IEEE Trans. Sustain. Energy **7**(1), 345–353 (2016)
13. Mendel, J.: Interval type-2 fuzzy logic systems: theory and design. IEEE Trans. Fuzzy Syst. **8**(5), 535e550 (2000)
14. Li, H.X.: A comparative design and tuning for conventional fuzzy control. IEEE Trans. Syst. Man Cybern. Part B Cybernet. Publication IEEE Syst. Man Cybernet. Soc. **27**(5), 884–9 (1997)

Tuning of Interval Type-2 Fuzzy Precompensated PID Controller: GWO-ABC Algorithm Based Constrained Optimization Approach

Prashant Gaidhane, Anupam Kumar, and Ritu Raj

Abstract With advances in technology, the control system becomes more complex and enhanced controllers are required for them. Fuzzy based controllers are always preferred for intelligent control of such systems. Fuzzy logic system is developed and advance versions are proposed. The interval type-2 fuzzy logic controller (IT2-FLC) has gained wide recognition for controlling systems with nonlinearities and uncertainties. This chapter presents systematic strategy to get maximum benefit from the shapes of the antecedent MF parameters. For experimental studies the interval type-2 fuzzy precompensated PID (IT2FP-PID) controller is designed for robotic arm and optimized for trajectory tracking problem. Various constraints are considered during optimization procedure. 60 parameters are tuned in a high dimensional problem using recent enhanced algorithm. The robustness of the controller in the presence of external disturbances, measurement noise and parameter variations is investigated. The results are compared with other equivalent counterparts. Minimization of performance metric integral time absolute error (ITAE) is selected as a objective function and hybrid grey wolf optimizer and artificial bee colony algorithm (GWO-ABC) is used.

P. Gaidhane (✉)
Government College of Engineering, Jalgaon, India
e-mail: pjgaidhane@gmail.com

A. Kumar
Department of Electronics and Communication Engineering, National Institute of Technology Patna, Patna, India
e-mail: anupam.ec@nitp.ac.in; anuanu1616@gmail.com

R. Raj
Department of Electronics and Communication Engineering, Indian Institute of Information Technology, Kota, India

© The Author(s), under exclusive license to Springer Nature Switzerland AG 2023
O. Castillo and A. Kumar (eds.), *Recent Trends on Type-2 Fuzzy Logic Systems: Theory, Methodology and Applications*, Studies in Fuzziness and Soft Computing 425,
https://doi.org/10.1007/978-3-031-26332-3_6

1   Introduction

In the last few decades, control system design has advanced by the incorporation of fuzzy logic systems (FLS). Thus, fuzzy logic controllers (FLC) proved their applicability in ill-structured nonlinear systems with uncertain parameters [26]. In literature, multiple characteristics of FLC are mentioned, like (i) low design, development and operating cost, (ii) rule base formation based on knowledge of human expertise, (iii) when exact mathematical model is unavailable, FLC can be applied based on dynamic model, and (iv) the ability of general framework of FLC to handle uncertainty, etc. Over the past few years, many authors proposed variants of FLC for different types of applications [13, 25]. Earlier, control designers mentioned FLC as type-1 fuzzy logic controller and recommended it over other conventional counterparts.

Pre-eminently, the concept of type-2 fuzzy set (IT2-FS) was proposed by Zadeh [40]. He extended the structure of type-1 fuzzy sets (T1-FS) by introducing uncertainty in IT2-FS. Subsequently, Karnik and Mendel [18], and other researchers [19] contributed a complete type-2 FLS theory by establishing characteristics of membership grades, theoretic working, and formulae for type-2 relation composition. To be specific, IT2-FLCs were upgraded with the rules assimilating antecedents or consequent with uncertainty therein [31]. Overall, the IT2-FLC, including the FOU in IT2-FSs, turn out as a developed FLC tool than T1-FLC.

Control system problems have larger uncertainties due to interference of unwanted signals in the closed-loop. T1-FLCs, with fuzziness or the uncertainty in between [0, 1], are found less suitable for complex control applications [8, 9]. After introduction of IT2-FLC, it has been seen as a viable alternative in such control problems. But, now the parameters of controllers are increased. On one side they provide better degree of freedom to designers, contrary, on the other side, it takes larger time for tuning methodologies to execute the operation.

In recent years, many researchers proposed different strategies to maximize the structures of IT2-FLC to extract maximum benefits from the FOU. Few studies [12, 28], have tested and presented strategies to rearrange structures of the IT2-FSs to get maximum advantage from them. Difficulty in tuning and high dimensional structural variables induce complexity and their limitations induces constraints in the optimization problem. As no exact mathematical model is available, Meta-heuristic optimization algorithm (MOEAs) are found perfect for optimization [35]. In last two decades, many researchers [8, 27, 31, 35] presented successful studies about optimizing IT2-FLC using MOEAs. Advancing on this research track, different MOEAs, namely, Tabu Search [5], ABC-GA [11, 26], Big Bang-Big Crunch optimization [28], cuckoo search [12, 36, 37], BFO [4], chaotic PSO [34], CSA optimization [1], etc., have been tested for variants of FLC on diverse applications.

Conventional PID controllers is observed to have limitations in robust control problem. FLC based precompensator PID (FP-PID) controllers was suggested in [22, 23] by Kim et al. The FLC induce many features like knowledge based inference, fuzzy decision making and enhance the results. In the response, unwanted undershoots and overshoots are minimized and steady state error is diminished.

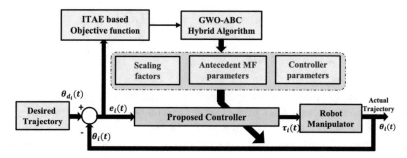

Fig. 1 Block diagram of overall control system design

Inspired by this, this chapter also present IT2-FLC based precompensated controller and further enhances the shapes of the antecedent MF structures (Fig. 1) [14].

The chapter contributes in terms of simple controller tuning approach using IT2-FLC and optimization algorithm. Several parameters of an advanced IT2FP-PID controller for a robotic manipulator with 2-DOF controls are optimized to track desired trajectory. Maximum benefit of IT2-FLC is acquired by optimizing shapes of antecedent MFs, while considering its constraints. This will also help to achieve optimal benefits of FOU in the proposed controller. If all parameters are considered, the optimization problem becomes high-dimensional constrained problem. Hence, the GWO-ABC algorithm is applied for its features and low computational cost.

The prime contributions of this chapter are given as follows:

- The benefits of IT2-FLC and the most of FOU are acquired using systematic strategy to tune the controller parameters, shape of MFs, and scaling factors.
- The implementation of interval type-2 fuzzy based precompensated PID controller is discussed for 2-DOF robotic manipulator's end-effectors path tracking problem.
- A technique to handle high-dimensional constrained optimization problem is described using hybrid GWO-ABC algorithm.
- The performance index ($ITAE$) is defined and the comprehensive analysis is carried out. The results are compared with other equivalent conventional controllers.
- The distinct nonlinear dynamics, like (a) signal disturbance, (b) feedback path noise, (c) model uncertainties, and (d) variations in payload are applied and the robustness analysis is presented.

Rest of the chapter progresses in this manner. Section 2 describes the essentials of GWO-ABC optimization algorithm and robotic manipulator. It also briefs the features of FLS structures of type-1 and type-2 FLCs. Proceeds by Sect. 3, defining detailed structure, implementation strategy, and tuning of IT2FP-PID controller. Simulation result analysis is discussed in the next section. Further, Sect. 5 covers robustness analysis for distinct nonlinear dynamics. Finally, the chapter is concluded in Sect. 6.

2   Optimization Algorithm and Type-2 FLC

2.1   GWO-ABC Algorithm

Several swarm intelligent based optimization algorithms were proposed by researchers in last few decades [33]. Few of them are widely appreciated for their ability to get globally optimized solution in available high dimensional search space. Then also, the conventional versions of these algorithms have certain limitations, such as, poor exploration abilities and immature locking to local optima [15]. Henceforth, with the aim to improve their performance, modified or hybrid versions are proposed by inculcating good methodologies of different algorithms. Novel GWO-ABC is such a newly proposed algorithm which inherit the advantages of conventional GWO [33] and ABC [17]. It is evident from the comparative analysis that GWO-ABC performs much better and have faster convergence.

The GWO-ABC succeeds in 3 phases, known as, population initialization phase, GWO phase, and ABC information sharing phase. The major features of the algorithm can be deciphered from [15]. By hybridizing major features of two algorithm, each solution element shares information with other element and overall global search technique is ameliorated to elude inappropriate stagnation. In controller tuning problems, with multiple- dimensional search space, such extensive techniques provide exhaustive coverage and attain jump-offs from any locally sub-optimal regions to deduce optimal controller parameters.

2.2   Features of Type-2 FLC

From the beginning of fuzzy sets, the problem of uncertainty handling is of concern. Contradicting to the nature of fuzziness, the uncertainty consideration is low in design of type-1 fuzzy set. In the basic type-1 fuzzy logic system (T1-FLS) structure, as depicted in Fig. 2a, different units like fuzzifier, inference engine, fuzzy rules, and defuzzifier are integrated. Initially, crisp input is transformed to type-1 fuzzy set (T1-FS) using fuzzifier and each input is designated by single layer MFs. We can observe from the triangular MFs in Fig. 2c that at any point (x') in T1-FS has unique MF value (μ'). Thus, there is no uncertainty about the value of the MFs and hence criticism was made. To' overcome the above criticism, more sophisticated type of fuzzy sets, i. e. "type-2 fuzzy set" incorporating uncertainty about their MFs were proposed by Lotfi A. Zadeh.

Figure 2b shows the structure of an interval type-2 fuzzy logic system (IT2-FLS). It can be observed that it is similar to T1-FLS shown in Fig. 2a. All units of FLC, (i) fuzzifier, (ii) inference engine, (iii) defuzzifier are common in both the FLS, extra unit (iv) type reducer is introduced.

The input signals to fuzzifier are converted to IT2-FS. These input signals may have uncertainties in the form of external noise and parameter variations. The

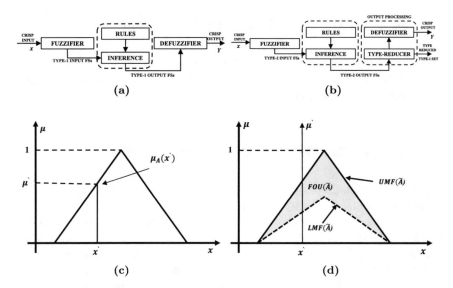

Fig. 2 General Illustrations of the **a** Structure of Type-1 fuzzy logic system, **b** Structure of Type-2 fuzzy logic system, **c** T1-FS, **d** IT2-FS

Table 1 The rule base of the type-1 and interval type-2 fuzzy logic controllers

E ⇓/CE ⇒	N	Z	P
N	NL	NM	Z
Z	NM	Z	PM
P	Z	PM	PL

IT2-FS represent the uncertainty in terms of finite region represented as *footprint of uncertainty* (FOU). As depicted in vertical slice representation in Fig. 2d, the IT2-FS is designated as UMF($\tilde{A}$)-the upper membership function, and LMF($\tilde{A}$)-lower membership function and noted as $\underline{\mu}_{\tilde{A}}(x), \forall x \in X$ and $\overline{\mu}_{\tilde{A}}(x), \forall x \in X$, respectively. For ease, the upper membership function is fixed to unity.

The operation of 'fuzzifier' is to map crisp input to type-2 fuzzy sets. Further, the inference engine is activated and output is generated. Type reducer converts the output signal to T1-FSs. Finally crisp output is obtained through defuzzifier. In this work, Karnik-Mendel (KM) algorithm is implemented among many in literature. Takagi-Sugeno-Kang (TSK) type IT2-FLSs can be obtained in the open source toolbox [38]. Details of type-2 FLC and reasoning mathematically of IT2-FS can be found in [19, 38, 39] (Table 1).

3   Controller Design and Optimization Method

The controller details are discussed in this section followed by MF structure representation and the strategy of MF optimization. At last, optimization algorithm implementation technique is presented.

3.1   Design of the Proposed IT2FP-PID Controller

Figure 3 demonstrates proposed IT2FP-PID controller, which is designed by connecting two modules (a) and (b) in series.

The first module-(a) is an interval type-2 (IT2) FLC based precompensated controller and the second module-(b) is a conventional PID controller. Module-(a) stabilizes the transient response by compensating unwanted overshoots and undershoots. Along with this, steady state error is diminished as per performance criteria. FLC in this module can be adjusted by incorporating the problem oriented knowledge based rule base [32]. The IT2-FLC have two inputs, symbolized as (1) $e(t)$—normalized error and (2) $\Delta e(t)$—rate of change of error. As shown, gain elements K_E and K_{CE} are multiplied to these inputs, respectively, as given below.

$$E(t) = K_E e(t) = K_E(Y_d(t) - Y(t)) \tag{1}$$

$$CE(t) = K_{CE}\frac{de(t)}{dt} = K_{CE}\Delta e(t) \tag{2}$$

where, $e(t)$ is obtained by taking difference between the actual and desired trajectory. The output of module-(a) U is multiplied by gain K_U to get $u(t)$ as shown:

$$u(t) = K_U \, IT2FLC \, (E(t), CE(t)) \tag{3}$$

here, the $IT2FLC \, (E, CE)$ is meant to be a nonlinear function based on inference of IT2-FLC;

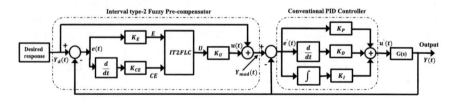

Fig. 3 Design of IT2 fuzzy precompensated PID controller

Thus, desired output for the first module obtained as

$$Y_{mod}(t) = u(t) + Y_d(t) \tag{4}$$

The new error signal $e'(t)$ is obtained as a difference between $Y_{mod}(t)$-the output response and $Y(t)$-the current response.

$$e'(t) = Y_{mod}(t) - Y(t) \tag{5}$$

This signal is fed to PID controller i.e. module-(b). The final actuating signal $u'(t)$ which moves the position of links, is derived as

$$u'(t) = K_P e'(t) + K_D \frac{de'(t)}{dt} + K_I \int e'(t)dt \tag{6}$$

3.2 Proposed Methodology for Optimization of IT2FP-PID Controller

IT2FP-PID controller is tuned to optimal values using systematic methodology presented in this section. Along with controller parameters, elements of antecedent MFs are also optimally reshaped to minimize the objective function.

Referring to Fig. 3, both the inputs to FLC—$e(t)$ and $\Delta e(t)$—are represented by a pair of three triangular MFs and output by five singletons at $1, 0.8, 0, -0.8$, and -1. The MFs are represented by different labels like 'P, Z, N' and 'PB, PM, Z, NM, NB' as shown in Fig. 6. In this work, the knowledge of designer is acquired in terms of 3×3 rule base. The details of MFs and FLC used in this work are represented as follows.

Input MFs Positive-P, Zero-Z, Negative-N.
Output MFs Positive Big-PB, Positive Medium-PM, Zero-Z, Negative Medium-NM, Negative Big-NB,
Each Type-1 Input MFs Triangular with labels $(l_{im}, c_{im}, r_{im}; i = 1, 2, \ldots$ and $m = 1, 2, 3. \ldots)$.
Each Type-2 Input MFs Triangular with labels $(l_{im}, c_{im}, r_{im}; i = 1, 2, \ldots$ and $m = 1, 2, 3. \ldots)$.
Rule base 3x3 rule base and Sugeno inference mechanism.
Software Fuzzy Logic Toolbox in MATLAB.

While optimizing the shapes of the MFs of both type-1 and type-2 FLC, few constraints are assigned and fulfilled to get new optimized shapes without losing the generosity of the classical FLC MFs. In this work we have 2 inputs $i = 2$ and 3 MFs for each input $m = 3$. For every triangular shape notation l is used for left corner, c is used for center and r is used to represent right corner. Thus, for type-1 input, we can name l_{1N}, c_{1N}, r_{1N} for N MF of input 1, and so for others. For type-2, we can name $l_{1N}, c_{1N}, r_{1N}, h_{1NU}, h_{1NL}$ for N MF of input 1, and so for others. All these labels are represented in Fig. 6.

To maintain the shape originality, following constraints are defined for input 1 and 2 of both type-1 and type-2 FLC

$$Input1 : c_{1N} < c_{1Z} < c_{1P}; l_{1N} < c_{1N} < r_{1N}; l_{1Z} < c_{1Z} < r_{1Z}, l_{1P} < c_{1P} < r_{1P}.$$

$$Input2 : c_{21N} < c_{2Z} < c_{2P}; l_{2N} < c_{2N} < r_{2N}; l_{2Z} < c_{2Z} < r_{2Z}, l_{2P} < c_{2P} < r_{2P}.$$

Along with this, we have to optimize the height of IT2-FSs. Hence, upper height is fixed to 1 and lower height is varied in between 0 to 1. The constraints are as given below

$$Input1 : 0 < h_{1NL} < 1; 0 < h_{1ZL} < 1; 0 < h_{1PL} < 1; h_{1NU} = h_{1ZU} = h_{1PU} = 1$$

$$Input2 : 0 < h_{2NL} < 1; 0 < h_{2ZL} < 1; 0 < h_{2PL} < 1; h_{2NU} = h_{2ZU} = h_{2PU} = 1$$

The parameters given above can be observed from Fig. 6a and b.

3.3   The Controller Implementation on Robotic Manipulator

Many studies reported that the control problem of MIMO machine robotic manipulator is nonlinear and coupled with other parameters. In this control problem, the torque (τ_1, τ_2) are manipulated based on the link positions (θ_1, θ_2). Figure 5 illustrate the input output control parameters used in this problem. Here, two links of robotic arm are regulated by two independent controllers employed at separate link. The 2 DOF robotic model and the technique is taken from [14] and the load is varied for further simulations. The parameters of the robotic manipulator used for the simulation study are listed in Table 2.

Controller implementation procedure and their parameters are explained here.

PID: Classical PID controller is used here. It is similar to module-(b) of IT2FP-PID controller as depicted in Fig. 3. Here, no FLC is used and only 3 controller parameters, namely K_P, K_I, and K_D are designated. For two such controllers attached to 2 links, 6 parameters are needed to tune.

FPID: FLC based PID controller, designed as FPID controller and depicted in Fig. 4. For both the controllers, 8 parameters to tune, namely K_{E_1}, K_{E_2}, K_{CE_1}, K_{CE_2}, K_{P_1}, K_{P_2}, K_{I_1}, and K_{I_2}. Basic triangular T1-MFs, as illustrated in Fig. 6a are implemented and 3x3 rule base is used.

TIFP-PID: The FPID controller structure is depicted in Fig. 5. Controller parameters K_E, K_{CE}, K_U, K_P, K_I, and K_D are to be optimized. Along with this, $9 \times 2 = 18$ variables used to define MFs are needed to tune per link. Thus for 2 separate links $2 \times (6 + 18) = 48$ parameters are required. As two inputs (E, CE) are applied to FLC and each input is represented by 3 MFs ('N', 'Z', and 'P'), their struc-

Table 2 Parameters of the robotic manipulator

Variables (unit)	Description	Link 1	Link 2
m_1, m_2	Mass (kg)	1	1
l_1, l_2	Length (m)	1	1
I_1, I_2	Centroid inertia (kgm^2)	0.2	0.2
v_1, v_2	Coefficient of viscous friction	0.1	0.1
d_1, d_2	Coefficient of dynamic friction	0.1	0.1
g	Acceleration due to gravity (m/s^2)	9.81	9.81
l_{c1}, l_{c2}	Distance between joint of link and center of gravity (m)	0.5	0.5
m_{pl}	The payload mass at the end of Link 2 (kg)	–	2 at $t = 0$

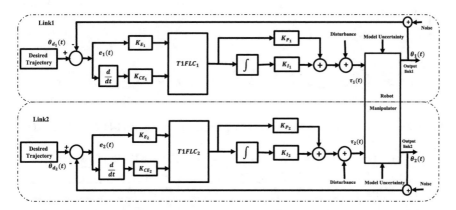

Fig. 4 Block diagram of FPID controller for 2—DOF robotic manipulator

tural parameters are designated as shown in Fig. 6a. Thus parameters $l_{1N}, c_{1N}, r_{1N}, l_{1Z}, c_{1Z}, r_{1Z}$ and l_{1P}, c_{1P}, r_{1P} are designated for link1. Similarly, parameters $l_{2N}, c_{2N}, r_{2N}, l_{2Z}, c_{2Z}, r_{2Z}$ and l_{2P}, c_{2P}, r_{2P} are designated for link2.

IT2FP-PID: As depicted in Fig. 5, the IT2FP-PID controller have similar structure as of FPID controller. Here also, controller parameters $K_E, K_{CE}, K_U, K_P, K_I,$ and K_D are to be optimized. Along with this, 24 variables used to define type-2 MFs are needed to tune per link. Thus for 2 separate links $2 \times (6 + 24) = 60$ parameters are required. In this case also, two inputs (E, CE) are applied to FLC and each input is represented by 3 MFs ('N', 'Z', and 'P'), their structural parameters are designated as shown in Fig. 6b. Thus parameters $l_{1N}, c_{1N}, r_{1N}, h_{1N}, l_{1Z}, c_{1Z}, r_{1Z}, h_{1Z}$ and $l_{1P}, c_{1P}, r_{1P}, h_{1P}$ are designated for link1. Similarly, parameters $l_{2N}, c_{2N}, r_{2N}, h_{2N}, l_{2Z}, c_{2Z}, r_{2Z}, h_{2Z}$ and $l_{2P}, c_{2P}, r_{2P}, h_{2P}$ are designated for link2.

Being high-dimensional optimization problem, GWO-ABC algorithm is applied for solving the problem. As reported in Table 3, all parameters are tuned to get minimum value of the objective function.

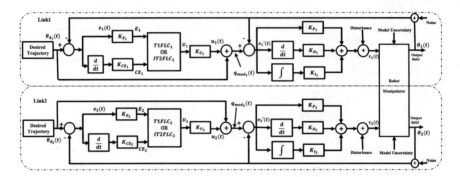

Fig. 5 Block diagram of T1FP-PID and proposed IT2FP-PID controllers for 2—DOF robotic manipulator

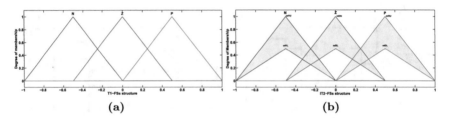

Fig. 6 General structures of input MFs for all the inputs and links of **a** T1-FLC and **b** IT2-FLC

Table 3 List of tuning parameters of the PID, FPID, T1FP-PID, and IT2FP-PID controllers

Contr.	Links	Antecedent MF parameters	Scaling factors
PID	L1	–	$K_{P_1}, K_{I_1}, K_{D_1}$
	L2	–	$K_{P_2}, K_{I_2}, K_{D_2}$
FPID	L1-I1/I2	Fixed Triangular MFs	$K_{E_1}, K_{CE_1}, K_{P_1}, K_{I_1}$
	L2-I1/I2		$K_{E_2}, K_{CE_2}, K_{P_2}, K_{I_2}$
T1FP-PID	L1-I1	$l_{1N_1}, c_{1N_1}, r_{1N_1}, l_{1Z_1}, c_{1Z_1}, r_{1Z_1},$ $l_{1P_1}, c_{1P_1}, r_{1P_1}$	$K_{E_1}, K_{CE_1}, K_{U_1}, K_{P_1}, K_{I_1}, K_{D_1}$
	L1-I2	$l_{1N_2}, c_{1N_2}, r_{1N_2}, l_{1Z_2}, c_{1Z_2}, r_{1Z_2},$ $l_{1P_2}, c_{1P_2}, r_{1P_2}$	
	L2-I1	$l_{2N_1}, c_{2N_1}, r_{2N_1}, l_{2Z_1}, c_{2Z_1}, r_{2Z_1},$ $l_{2P_1}, c_{2P_1}, r_{2P_1}$	$K_{E_2}, K_{CE_2}, K_{U_2}, K_{P_2}, K_{I_2}, K_{D_2}$
	L2-I2	$l_{2N_2}, c_{2N_2}, r_{2N_2}, l_{2Z_2}, c_{2Z_2}, r_{2Z_2},$ $l_{2P_2}, c_{2P_2}, r_{2P_2}$	
IT2FP-PID	L1-I1	$l_{1N_1}, c_{1N_1}, r_{1N_1}, h_{1N_1}, l_{1Z_1}, c_{1Z_1},$ $r_{1Z_1}, h_{1Z_1} l_{1P_1}, c_{1P_1}, r_{1P_1}, h_{1P_1}$	$K_{E_1}, K_{CE_1}, K_{U_1}, K_{P_1}, K_{I_1}, K_{D_1}$
	L1-I2	$l_{1N_2}, c_{1N_2}, r_{1N_2}, h_{1N_2}, l_{1Z_2}, c_{1Z_2},$ $r_{1Z_2}, h_{1Z_2} l_{1P_2}, c_{1P_2}, r_{1P_2}, h_{1P_2}$	
	L2-I1	$l_{2N_1}, c_{2N_1}, r_{2N_1}, h_{2N_1}, l_{2Z_1}, c_{2Z_1},$ $r_{2Z_1}, h_{2Z_1} l_{2P_1}, c_{2P_1}, r_{2P_1}, h_{2P_1}$	$K_{E_2}, K_{CE_2}, K_{U_2}, K_{P_2}, K_{I_2}, K_{D_2}$
	L2-I2	$l_{2N_2}, c_{2N_2}, r_{2N_2}, h_{2N_2}, l_{2Z_2}, c_{2Z_2},$ $r_{2Z_2}, h_{2Z_2} l_{2P_2}, c_{2P_2}, r_{2P_2}, h_{2P_2}$	

I1:- Input1 '$e(t)$', I2:-Input2 '$\Delta e(t)$'

4   Simulation Results and Analysis

This section presents the performance analysis of IT2FP-PID controller with classical PID, Fuzzy PID , and T1FP-PID controllers. As discussed above in Figs. 4 and 5, individual controllers are applied to every link and tuning parameters required for these controllers are listed in Table 3.

4.1   Deriving Objective Function

Proper selection of objective function is important in optimization problem. In this simulation, minimization of integral time absolute error ($ITAE$) is opted for its suitable features. It is found reliable for least percentage of overshoot ($\%M_p$) and reduction of oscillations. The responses settling time (t_s) and rise time (t_r) are also found to be reduced, which eventually minimize the steady state error E_{ss}. Thus, for error $e(t)$ over time t, the objective function is formulated as

$$Obj\_fun = ITAE = \int t|e(t)|dt \tag{7}$$

For two links of manipulator with errors e_{l1} and e_{l2}, the objective functions are reformed as

$$Obj\_fun_{l1} = ITAE_1 = \int t|e_{l1}(t)|dt \tag{8}$$

$$Obj\_fun_{l2} = ITAE_2 = \int t|e_{l2}(t)|dt \tag{9}$$

Single objective function is designed by taking weighted sum, with $w_1 = 1$ and $w_2 = 1$, of both link functions.

$$Obj\_fun = w_1 \times Obj\_fun_{l1} + w_2 \times Obj\_fun_{l2} \tag{10}$$

4.2   Experimental Settings

The experimental setup for simulation studies is uniform for different controllers. Some parameter ranges for links L_i where $i = 1, 2$, are

Algoritham setup   30 search agents, 100 iterations, 20 runs.
$K_{E_i}, K_{CE_i}, K_{U_i}$ Range(L_b, U_b) [0,500]
$K_{U_i}, K_{P_i}, K_{I_i}, K_{D_i}$ Range(L_b, U_b) [0,500]
Software   MATLAB Simulink, fuzzy logic toolbox by MATLAB [38].

4.3   Results and Discussion

The initial shapes of MFs for these FPID, T1FP-PID, and IT2FP-PID controllers are represented in Fig. 6a and b, respectively. From these shapes we can decide the Min-Max values of structural parameters to define constraints, as given above. After complete experimentation and simulation of different controllers the optimized scaling parameters are noted in Table 4 and corresponding optimized MFs structures are depicted in Figs. 7 and 8, for T1FP-PID and IT2FP-PID controllers, respectively. The most remarkable observation to emerge from these plots is that the optimized structures maintained the conditions of constrains and the triangular shapes, only, their width is varied to get most of FOU.

The values in Table 4 are noted when simulation of all iterations are over, i.e. after completion of stopping criterion. The values for IT2FP-PID for both the links

Table 4 Table showing results of parameter values of controller after optimization

Controllers ⇒	PID		FPID		T1FP-PID		IT2FP-PID	
Parameter ⇓	Link1	Link2	Link1	Link2	Link1	Link2	Link1	Link2
K_P	423.02	425.11	125.71	16.121	401.11	412.11	17.121	186.02
K_I	97.541	1.0121	131.062	0.0054	79.441	41.041	425.31	8.5460
K_D	145.40	477.11	–	–	0.0452	0.3115	0.0014	0.0024
K_E	–	–	22.488	14.481	55.785	438.41	478.31	221.27
K_{CE}	–	–	0.1745	1.0010	0.0004	0.0049	0.0521	0.0021
K_U	–	–	–	–	98.527	2.8455	69.760	21.8651
$ITAE$	0.08240	0.0612	0.0112	0.0114	0.002142	0.001249	**0.00000121**	**0.00000341**

The "best results" are indicated by bold values

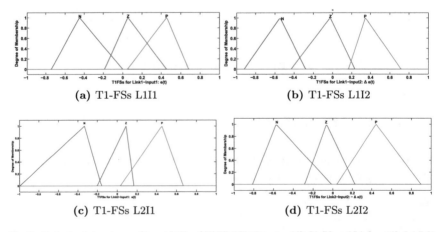

(a) T1-FSs L1I1

(b) T1-FSs L1I2

(c) T1-FSs L2I1

(d) T1-FSs L2I2

Fig. 7 Optimized structures of input MFs of T1FP-PID (Input—1/2: I1 /I2 and Link—1/2: L1/L2)

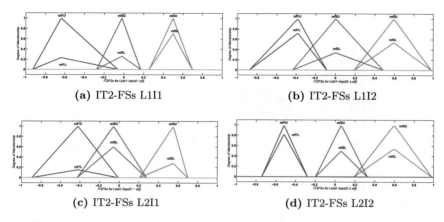

(a) IT2-FSs L1I1

(b) IT2-FSs L1I2

(c) IT2-FSs L2I1

(d) IT2-FSs L2I2

Fig. 8 Optimized structures of input MFs of IT2FP-PID (Input—1/2: I1 /I2 and Link—1/2: L1/L2)

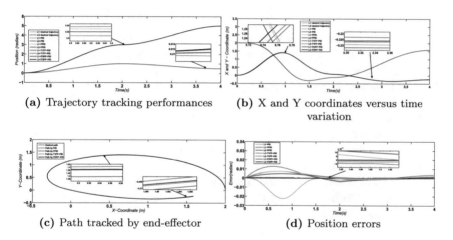

(a) Trajectory tracking performances

(b) X and Y coordinates versus time variation

(c) Path tracked by end-effector

(d) Position errors

Fig. 9 Comparative analysis of responses of robotic end-effector with variable payload by optimized controllers

are much smaller than other controllers values. The comparative analysis of $ITAE$ values, 2.41×10^{-06} and 5.76×10^{-06}, for the separate links of IT2FP-PID, are also demonstrated in Fig. 10. The least values substantiate the excellence of IT2FP-PID controller and ensure the exact tracking of defined desired path by tip of the robotic arm. Finally, it is also evident that the GWO-ABC algorithm accomplishes optimal values of performance metric after constrained optimization.

To exhibit further proofs of performance of proposed controller, various plots of different nature are demonstrated. In these plots desired results are demonstrated in comparison with results obtained by all four IT2FP-FPID, T1FP-PID, FPID, and PID controllers. For further convenience, the enlarged illustrations are also included in the plot to get better view. These plots and observations are noted as follows.

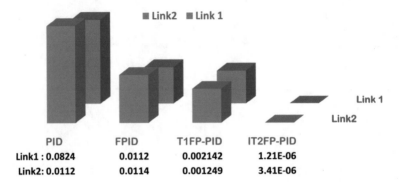

Fig. 10 Analysis of minimized ITAE values for all controllers applied to Link1 and Link2 (Not to the scale)

1. Figure 9a shows the desired trajectory tracking by both the robotic links. Exact path tracking of IT2FP-PID can be observed as it effectively eradicates the overshoots and undershoots.
2. Figure 9b demonstrates X and Y coordinate of tip of the robotic arm versus time taken by the arm. Optimized IT2FP-PID follows both the coordinates in an efficient way and accordingly enhance the performance.
3. Figure 9c demonstrates path traced by robotic end effector with respect to the predefined desired trajectory. Thus, in case of IT2FP-PID, the path following is more accurate than others.
4. Figure 9d demonstrates position errors for both the links generated by comparing actual results with desired results. We can clearly observe how IT2FP-PID reduces the position errors from initial time.

5   Robustness Analysis

In control system design procedures, along with desired response analysis, robustness analysis for different variations like

1. Model uncertainties i.e. variations in system parameters,
2. Low frequency external disturbances in the control path,
3. High frequency random feedback noise

are absolutely essential. Hence, the efficiency of IT2FP-PID is again tested after inserting following non-linearity in the system under unusual circumstances. These signals are introduced in the forward and feedback path as depicted in Figs. 4 and 5. The experimentation and simulation procedure described earlier is followed for these non-linearity, as well, Thus, all controllers are tuned to minimize $ITAE$. Lets discuss the robustness analysis after introducing different variations one-by-one in the following sections.

5.1 Analysis of Impact of Model Uncertainties

Timely changes in physical properties of robotic structure may vary its proper behaviour and affect the overall performance. Some variations such as frictional forces affects the mathematical model and balance of robotic manipulator The lubricants in the joint and other timely wear and tear affects these frictional forces [10]. Frictions like Coulomb friction and viscous friction may also degrade the operation. Henceforth, robustness against such periodic changes must have to be tackled by applied controllers. In this study, we apply $\pm 5\%$ and / or $\pm 20\%$ variations to different physical characteristics, as given below,

$\pm 5\%$:-   masses of links i.e. m_1, m_2, and $m_1 + m_2$
$\pm 20\%$:- lengths of the links $L1, L2$ i.e. l_1, l_2 and $l_1 + l_2$
$\pm 20\%$:- coefficient of viscous friction i.e. v_1, v_2 and $v_1 + v_2$
$\pm 20\%$:- coefficient of dynamic friction i.e. d_1, d_2 and $d_1 + d_2$

Thus, these physical characteristics are increased or decreased, as above, and the values of objective function are recorded. Table 5 noted the individual and collective results for these variations. At last, the overall average percentage variation with respect to original $ITAE$ values are calculated and noted in last row. From Table 5, it is evident that the IT2FP-PID controller results are varied by negligibly small value 1.21%. Whereas, other controllers performance like PID (7.128%), FPID (4.9124%), and T1FP-PID (4.965%) is varied by considerably larger values which affect their robustness against the physical parameter variations. Ipso facto, the overall findings clarify the superiority of IT2FP-PID over others and authenticate the impression of FOU in MFs to ensure the boundedness of the position tracking error.

5.2 Analysis of Impact of External Disturbances

Influences of external disturbance always affects the devices, equipment, instruments, and machines in industrial process control system. Hence, analysis of impact of external disturbances is very essential for any controller. As illustrated in Figs. 4 and 5, the disturbances are incorporated in respective links and control output is examined. Distinct disturbance signals of the nature '$A sin \omega t$' N-m, is applied to both the links, as represented in table shown in Fig. 6. The simulation is carried out for time period of 4s. The results of objective function for all the controllers are examined and noted in Table 6. Amplitude and frequency of the disturbance signal are varied for wider analysis.

Table 5 Results for analysis of impact of model uncertainties

Controllers ⇒	PID		FPID		T1FP-PID		IT2FP-PID	
Parameter ⇓	Link1	Link2	Link1	Link2	Link1	Link2	Link1	Link2
Optimum $ITAE$	0.08220	0.01210	0.01837	0.01414	0.00217	0.000121	0.00000195	0.00000475
m_1 (−5%)	0.08610	0.02031	0.01801	0.01242	0.00369	0.000172	0.00000241	0.00000576
m_2(−5%)	0.08620	0.02053	0.01811	0.01201	0.00329	0.000161	0.00000240	0.00000571
$m_1 + m_2$ (−5%)	0.08630	0.02041	0.01836	0.01205	0.00366	0.000171	0.00000238	0.00000571
l_1 (−5%)	0.08779	0.02393	0.01795	0.01357	0.00381	0.000162	0.00000238	0.00000575
l_2 (−5%)	0.08711	0.02148	0.01725	0.01186	0.00560	0.000124	0.00000274	0.00000555
$l_1 + l_2$ (−5%)	0.08757	0.02124	0.01785	0.01265	0.00247	0.000144	0.00000247	0.00000574
All m and l (−5%)	0.07985	0.01935	0.01784	0.01142	0.00245	0.000154	0.00000244	0.00000574
$v_1 + v_2$ (−15%)	0.0824	0.02444	0.01245	0.01124	0.00214	0.000144	0.00000212	0.00000457
$d_1 + d_2$ (−15%)	0.09134	0.02155	0.01939	0.01324	0.00354	0.000144	0.00000212	0.00000457
All d and v (−15%)	0.09137	0.02175	0.01930	0.01337	0.00373	0.000182	0.00000214	0.00000746
All above	0.04772	0.017545	0.02459	0.01774	0.00245	0.000145	0.00000144	0.00000541
m_1 (+5%)	0.08811	0.02888	0.01470	0.01551	0.00345	0.000145	0.00000124	0.00000457
m_2 (+5%)	0.08547	0.02124	0.01642	0.01124	0.00124	0.000124	0.00000124	0.00000452
$m_1 + m_2$ (+5%)	0.03454	0.04421	0.01142	0.01575	0.00127	0.000128	0.00000232	0.00000451
l_1 (+5%)	0.0451	0.01244	0.01245	0.01440	0.00441	0.000175	0.00000285	0.00000565
l_2 (+5%)	0.02540	0.02122	0.02124	0.01245	0.00541	0.000144	0.00000124	0.00000452
$l_1 + l_2$ (+5%)	0.09451	0.02443	0.02002	0.01355	0.00374	0.000156	0.00000124	0.00000242
All m and l (+5%)	0.08534	0.02124	0.01245	0.01744	0.00344	0.000147	0.00000742	0.00000451
$v_1 + v_2$ (+15%)	0.0865	0.02659	0.01456	0.01231	0.00451	0.000145	0.00000124	0.00000545
$d_1 + d_2$ (+15%)	0.09108	0.02122	0.01935	0.01305	0.00457	0.000156	0.00000145	0.00000457
All v and d (+15%)	0.0254	0.0124	0.014	0.01294	0.00364	0.000166	0.00000223	0.00000527
All above	0.09786	0.02024	0.04104	0.01347	0.00347	0.000144	0.00000122	0.00000154
Overall % variation	7.128%		4.9124%		4.965%		**1.21%**	

The standard deviation (STD) of $ITAE$ for separate links are observed for the given controllers. IT2FP-PID controllers have least deviation even after the introduction of various external disturbances. The results are represented in bold. Thus, proposed controller perform efficiently against disturbance variations compared to other controllers. For graphical representation, disturbance of $2.0sin60t$ N-m is applied in both the links of robotic arm. These plots and observations are noted as follows. For further convenience, the enlarged illustrations are also included in the plot to get better view.

1. Figure 11a shows the desired trajectory tracking by both the robotic links.
2. Figure 11b demonstrates X and Y coordinate of tip of the robotic arm versus time taken by the arm.
3. Figure 11c demonstrates path traced by robotic end effector with respect to the predefined desired trajectory.
4. Figure 11d demonstrates position errors for both the links generated by comparing actual results with desired results.

The STD values obtained from all the controllers demonstrates that IT2FP-PID gives the best performance in the presence of external disturbances.

5.3 Analysis of Impact of Random Feedback Noise

As demonstrated in Figs. 4 and 5, sensors are connected in feedback path in closed loop control. Generally, high frequency random noise is induced in the system through these sensors, because of some environmental conditions, and some repercussions are added in overall system due to these signals. This section conducted a comprehensive experimentation for analysis of impact of random feedback noise. In this view, random noise is applied in feedback path of either or both Link 1 and Link 2 and different plots are obtained as discussed below,

1. Figure 12a demonstrates the high-frequency random noise signal applied to feedback path in both the links.
2. Figure 12b demonstrates the trajectory tracking performance by both the robotic links.
3. Figure 12c demonstrates X and Y coordinate of tip of the robotic arm versus time variation.
4. Figure 12d demonstrates path tracked by robotic end effector with respect to the given desired trajectory.
5. Figure 12e demonstrates position errors for both the links generated by comparing actual results with desired results.

Minimized $ITAE$ values are noted in Table 7. Finally, all results drive us to conclude that the controller and optimized shapes enhance the performance of the robotic arm. The optimized IT2-FS, depicted in Fig. 8, in IT2FP-PID also helps to suppress the external noise and restrain uncertainties in the system compared to other

Table 6 Results for analysis of impact of external disturbances

Controllers ⇒		PID		FPID		T1FP-PID		IT2FP-PID	
Disturbance ⇓		Link1	Link2	Link1	Link2	Link1	Link2	Link1	Link2
Link1	$1sin60t$	0.08641	0.02145	0.01875	0.01345	0.00245	0.000164	0.00000325	0.00000457
	$2sin50t$	0.08452	0.01886	0.01754	0.01457	0.00299	0.000178	0.00000345	0.00000475
	$3sin50t$	0.0854	0.02475	0.01895	0.01345	0.00386	0.000177	0.00000344	0.00000457
	$1sin30t$	0.08457	0.01243	0.01754	0.01112	0.00124	0.000174	0.00000344	0.00000457
	$2sin30t$	0.04575	0.01244	0.0124	0.0454	0.00245	0.000144	0.00000244	0.00000452
	$3sin30t$	0.0245	0.0445	0.01244	0.0114	0.00454	0.000111	0.00000314	0.00000452
	STD	3.24E-4	7.12E-4	7.11E-4	3.74E-4	2.59E-5	8.44E-6	**1.14E-07**	**1.74E-08**
Link2	$1sin60t$	0.09141	0.02447	0.01247	0.01245	0.00246	0.000174	0.00000124	0.00000456
	$2sin60t$	0.0864	0.02457	0.01754	0.01234	0.00454	0.000144	0.00000124	0.00000457
	$3sin60t$	0.0854	0.01245	0.01774	0.01554	0.00444	0.000452	0.0000063	0.00000523
	$1sin30t$	0.04566	0.0223	0.014427	0.01745	0.00241	0.000145	0.00000210	0.00000542
	$2sin30t$	0.02154	0.01241	0.01124	0.01241	0.00121	0.000121	0.00000134	0.00000541
	$3sin30t$	0.01245	0.02414	0.01745	0.01124	0.00211	0.000201	0.00000231	0.00000678
	STD	7.21E-6	1.04E-3	2.14E-4	1.41E-5	7.40E-5	3.26E-6	**6.34E-08**	**2.01E-07**
Both	$1sin60t$	0.04521	0.02114	0.01544	0.01124	0.00124	0.000158	0.00000203	0.00000785
	$2sin60t$	0.08451	0.02124	0.01745	0.01241	0.00750	0.000174	0.00000548	0.00000548
	$3sin60t$	0.07541	0.0120	0.01124	0.0124	0.00854	0.000201	0.00000108	0.00000304
	$1sin30t$	0.05421	0.01048	0.01541	0.01885	0.00341	0.000111	0.00000345	0.00000644
	$2sin30t$	0.09174	0.01241	0.01754	0.01452	0.00246	0.000174	0.00000311	0.00000654
	$3sin30t$	0.0854	0.02124	0.01447	0.01321	0.00301	0.000145	0.00000311	0.00000641
	STD	3.75E-4	9.44E-4	1.32E-4	5.01E-4	4.77E-5	4.05E-6	**9.44E-08**	**7.10E-08**

The "best results" are indicated by bold values

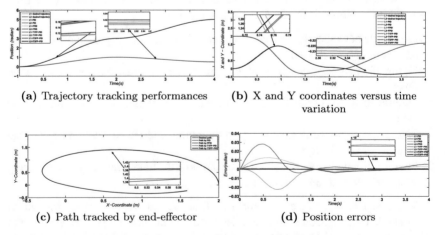

(a) Trajectory tracking performances

(b) X and Y coordinates versus time variation

(c) Path tracked by end-effector

(d) Position errors

Fig. 11 Comparative analysis of responses of robotic end-effector when external disturbance signal of $2.0sin60t$ is introduced in both the links

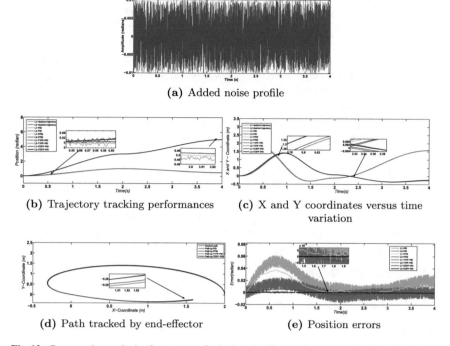

(a) Added noise profile

(b) Trajectory tracking performances

(c) X and Y coordinates versus time variation

(d) Path tracked by end-effector

(e) Position errors

Fig. 12 Comparative analysis of responses of robotic end-effector when external noise is introduced in both the links

Table 7 Results for analysis of impact of random feedback noise

Controllers ⇒	PID		FPID		T1FP-PID		IT2FP-PID	
Random noise at ⇓	Link1	Link2	Link1	Link2	Link1	Link2	Link1	Link2
Link1	0.1124	0.02654	0.03951	0.01021	0.00395	0.00586	0.0003865	0.0004124
Link2	0.1214	0.04547	0.04754	0.01124	0.003955	0.005784	0.0005112	0.0004125
Both links	0.10241	0.040214	0.04965	0.01021	0.00341	0.002354	0.0004562	0.0004325

controllers with un-optimized MFs. The response of classical PID and FPID, using uniform MFs, is not up to expectation and hence not suggested for handling complex systems with uncertainties.

6 Conclusions

Multiple observations are reported in different studies that IT2-FLC, with FOU in MFs, is manifested as an efficient tool for dealing with nonlinearities, disturbances, and high level of uncertainties in control system. However, the IT2-FLC based optimization problem is formulated as a complex high-dimensional constrained problem. Hence, overall parameter optimization takes more time to get best results. In this chapter, we proposed an efficient technique to tune the IT2FP-PID controller and its MFs structure. Constrained handling procedure is also proposed by selection of proper boundary. The controller is examined for trajectory tracking problem of 2-DOF robotic manipulator when applied with variable payload. In this experimentation, minimization of performance metric integral time absolute error ($ITAE$) is used as an objective function.

In interval type-2 fuzzy precompensated controller IT2-FLC is cascaded with basic PID controller. Additionally, the shapes of antecedent MFs are optimally designed to get best advantages of FOU. This enhancement compensates for the effect of unknown nonlinearities and reject the overshoots and undershoots. Recently introduced hybrid GWO-ABC algorithm is applied and results are evaluated against performance of T1FP-PID, FPID, and conventional PID controllers.

Comparative results and analysis in illustrations and tables significantly emphasize the effectiveness of IT2-FLC in the proposed controller. Robustness analysis for different nonlinear elements like

- Model uncertainties i.e. variations in system parameters,
- Low frequency external disturbances in the control path,
- High frequency random feedback noise.

has been carried out which emphasize the observations and support the positive effect of the FOU. Overall, the controller is robust and efficient compared to other counterparts. As a whole, following observations are noted

1. Maximum benefit of FOU can be derived from IT2-FLC with optimized shapes of MFs. It will enhance the performance to handle disturbances and uncertainty.
2. The proposed IT2FP-PID controller demonstrates its efficiency to control complex nonlinear systems with high uncertainties,
3. Added tuning parameters extend extra degree of freedom to the designer to enhance the performance.
4. Swarm inspired hybrid GWO-ABC algorithm revealed as viable alternative for the low- and high-dimensional constrained optimization problems.

As a extension to this work, experimental investigations of proposed controller and its real-time evaluation can be carried out. Design and implementation of hardware model of IT2FP-PID could be tried, which can provide an interesting opportunity for further research.

References

1. Aggarwal, A., Rawat, T.K., Upadhyay, D.K.: Design of optimal digital FIR filters using evolutionary and swarm optimization techniques. AEU - Int. J. Electron Commun. **70**, 373–85 (2016). https://doi.org/10.1016/j.aeue.2015.12.012
2. Alavandar, S., Jain, T., Nigam, M.J.: Bacterial foraging optimized hybrid fuzzy precompensated PD control of two link rigid-flexible manipulator. Int. J. Comput. Intell. Syst. **2**, 51–9 (2009). https://doi.org/10.2991/jnmp.2009.2.1.6
3. Angel, L., Viola, J.: Fractional order PID for tracking control of a parallel robotic manipulator type delta. ISA Trans. **79**, 172–88 (2018)
4. Arya, Y., Kumar, N.: A-scaled fractional order fuzzy PID controller applied to AGC of multi-area multi-source electric power generating systems. Swarm Evol. Comput. **32**, 2002–218 (2016). https://doi.org/10.1016/j.swevo.2016.08.002
5. Ate, A., Yeroglu, C.: Optimal fractional order PID design via Tabu search based algorithm. ISA Trans. **60**, 109–118 (2016)
6. Bosque, G., Del Campo, I., Echanobe, J.: Fuzzy systems, neural networks and neuro-fuzzy systems: a vision on their hardware implementation and platforms over two decades. Eng. Appl. Artif. Intell. **32**, 283–331 (2014)
7. Castillo, O., Melin, P.: Genetic optimization of interval type-2 fuzzy systems for hardware implementation on FPGAs. In: Recent Advances in Interval Type-2 Fuzzy Systems. Springer Briefs in Applied Sciences and Technology, vol. 1. Springer, Berlin (2012)
8. Castillo, O., Melin, P.: A review on interval type-2 fuzzy logic applications in intelligent control. Inf. Sci. (Ny) **279**, 615–31 (2014)
9. Castillo, O., Amador-Angulo, L., Castro, J.R., Garcia-Valdez, M.: A comparative study of type-1 fuzzy logic systems, interval type-2 fuzzy logic systems and generalized type-2 fuzzy logic systems in control problems. Inf. Sci. (Ny) **354**, 257–74 (2016). https://doi.org/10.1016/j.ins.2016.03.026
10. Craig, J.J.: Introduction to Robotics: Mechanics and Control, 2nd ed. Addison-Wesley Longman Publishing Co., Inc., Boston (1989)
11. Das, S., Pan, I., Das, S., Gupta, A.: A novel fractional order fuzzy PID controller and its optimal time domain tuning based on integral performance indices. Eng. Appl. Artif. Intell. **25**, 430–42 (2012). https://doi.org/10.1016/j.engappai.2011.10.004
12. Fatihu Hamza, M., Jen Yap, H., Ahmed, C.I.: Cuckoo search algorithm based design of interval Type-2 Fuzzy PID Controller for Furuta pendulum system. Eng. Appl. Artif. Intell. **62**, 134–51 (2017). https://doi.org/10.1016/j.engappai.2017.04.007
13. Gaidhane, P.J., Kumar, A., Nigam, M.: Tuning of two-DOF-FOPID controller for magnetic levitation system: a multi-objective optimization approach. In: 6th IEEE International Conference Computer Application Electrical Engineering - Recent Advances, pp. 497–502 (2017)
14. Gaidhane, P.J., Nigam, M.J., Kumar, A., Pradhan, P.M.: Design of interval type-2 fuzzy precompensated PID controller applied to two-DOF robotic manipulator with variable payload. In: ISA Transactions. Elsevier (2018)
15. Gaidhane, P.J., Nigam, M.J.: A hybrid grey wolf optimizer and artificial bee colony algorithm for enhancing the performance of complex systems. J. Comput. Sci. **27**, 284–302 (2018). https://doi.org/10.1016/j.jocs.2018.06.008
16. Hagras, H.: A hierarchical type-2 fuzzy logic control architecture for autonomous mobile robots. IEEE Trans. Fuzzy Syst. **12**, 524–39 (2004)
17. Karaboga, D., Basturk, B.: A powerful and efficient algorithm for numerical function optimization: artificial bee colony (ABC) algorithm. J. Glob. Optim. **39**, 459–71 (2007)
18. Karnik, N.N., Mendel, J.M.: Introduction to type-2 fuzzy logic systems. In: Proceeding IEEE FUZZ Conference, Anchorage (1998)
19. Karnik, N.N., Mendel, J.M., Liang, Q.: Type-2 fuzzy logic systems. IEEE Trans. Fuzzy Syst. **7**, 643–658 (1999)
20. Khosla, M., Sarin, R., Uddin, M.: Design of an analog CMOS based interval type-2 fuzzy logic controller chip. J. Artif. Intell. Expert. **2**, 167–183 (2011)

21. Kim, D.: An Implementation of fuzzy logic controller on the reconfigurable FPGA system. IEEE Trans. Ind. Electron. **47**, 703–715 (2000)
22. Kim, J., Park, J., Lee, S., Chong, E.K.P.: Fuzzy precompensation of PD controllers for systems with deadzones. J. Intell. Fuzzy Syst. **1**, 125–33 (1993)
23. Kim, J.H., Kim, K.C., Chong, E.K.P.: Fuzzy precompensated PID controllers. IEEE Trans. Control Syst. Technol. **2**, 406–11 (1994). https://doi.org/10.1109/87.338660
24. Kishor, A., Singh, P.K.: Empirical study of grey wolf optimizer. Adv. Intell. Syst. Comput. **436**, 1037–49 (2016). https://doi.org/10.1007/978-981-10-0448-3-87
25. Kumar, A., Gaidhane, P.J., Kumar, V.: A nonlinear fractional order pid controller applied to redundant robot manipulator. In: 6th IEEE International Conference Computer Application Electrical Engineering - Recent advances, pp. 545–550 (2017)
26. Kumar, A., Kumar, V.: Evolving an interval type-2 fuzzy PID controller for the redundant robotic manipulator. Expert Syst. Appl. **73**, 161–77 (2016). https://doi.org/10.1016/j.eswa. 2016.12.029
27. Kumar, A., Kumar, V.: A novel interval type-2 fractional order fuzzy PID controller: design, performance evaluation, and its optimal time domain tuning. ISA Trans. **68**, 251–75 (2017). https://doi.org/10.1016/j.isatra.2017.03.022
28. Kumbasar, T., Hagras, H.: Big Bang-Big Crunch optimization based interval type-2 fuzzy PID cascade controller design strategy. Inf. Sci. (Ny) **282**, 277–95 (2014). https://doi.org/10.1016/ j.ins.2014.06.005
29. Kumbasar, T., Hagras, H.: Interval Type-2 Fuzzy PID Controllers, pp. 285–94. Springer Handbook of Computational Intelligence, Berlin (2015)
30. Mendel, J.M., Hagras, H., John, R.I.: Standard background material about interval type-2 fuzzy logic systems that can be used by all authors. IEEE Comput. Intell. Soc. 1–11 (2010)
31. Mendel, J.M., John, R.I., Liu, F.: Interval type-2 fuzzy logic systems made simple. IEEE Trans. Fuzzy Syst. **14**, 808–21 (2006)
32. Meza, J.L., Santibez, V., Soto, R., Llama, M.A.: Fuzzy self-tuning PID semiglobal regulator for robot manipulators. IEEE Trans. Ind. Electron **59**, 2709–2717 (2012)
33. Mirjalili, S., Mirjalili, S.M., Lewis, A.: Grey Wolf optimizer. Adv. Eng. Softw. **69**, 46–61 (2014)
34. Pan, I., Das, S.: Fractional order fuzzy control of hybrid power system with renewable generation using chaotic PSO. ISA Trans. **62**, 19–29 (2016). https://doi.org/10.1016/j.isatra.2015. 03.003
35. Pan, I., Das, S., Gupta, A.: Tuning of an optimal fuzzy PID controller with stochastic algorithms for networked control systems with random time delay. ISA Trans. **50**, 28–36 (2011). https:// doi.org/10.1016/j.isatra.2010.10.005
36. Sharma, R., Gaur, P., Mittal, A.P.: Performance analysis of two-degree of freedom fractional order PID controllers for robotic manipulator with payload. ISA Trans. **58**, 279–91 (2015). https://doi.org/10.1016/j.isatra.2015.03.013
37. Sharma, R., Gaur, P., Mittal, A.P.: Design of two-layered fractional order fuzzy logic controllers applied to robotic manipulator with variable payload. Appl. Soft Comput. J. **47**, 565–76 (2016). https://doi.org/10.1016/j.asoc.2016.05.043
38. Taskin, A., Kumbasar, T.: An open source Matlab/simulink toolbox for interval type-2 fuzzy logic systems. Comput. Intell 2015 IEEE Symp. Ser. 2015, 1561–1568. https://doi.org/10. 1109/SSCI.2015.220
39. Wu, D., Mendel, J.M.: Enhanced Karnik - Mendel algorithms. IEEE Trans. Fuzzy Syst. **17**, 923–34 (2009)
40. Zadeh, L.A.: The concept of a linguistic variable and its application to approximate reasoning-I. Inform. Sci. **8**, 199–249 (1975)

Design of Type 2 Fuzzy Controller for OWC Power Plant

Sunil Kumar Mishra, Mano Ranjan Kumar, Bhargav Appasani, Amitkumar Vidyakant Jha, and Avadh Pati

Abstract The oscillating water column (OWC) wave power plants' rotor speed control is suggested in this book chapter. To get the most electrical output power from ocean waves, a Type 2 Fuzzy Logic Controller (T2FLC) is used to manage the rotational speed. A maximum power point tracking (MPPT) method is created that offers the best rotor speed reference to do this. The main goal of T2FLC is to match the reference speed with the real rotor speed. For input fuzzification, a triangle membership function of the interval type has been selected, and an inference engine of the Takagi–Sugeno type has been suggested. T2FLC's output is considered to be of the crisp type. Finally, the realistic JONSWAP ocean wave model has been taken into account while doing the extensive simulations. The suggested T2FLC has been used to examine important OWC plant metrics such turbine power, output power, turbine flow coefficient, and rotor speed.

Keywords Ocean energy · Oscillating water column · Rotation speed control · Power generated · Type 2 fuzzy logic control

1 Introduction

Ocean energy is a form of renewable energy that may be used in a variety of ways to produce electricity. When converting wave energy into electrical energy, oscillating water columns (OWC) are often employed techniques. The control system plays crucial role in the proper functioning of the OWC plants. Several control approaches have been applied in the control of OWC plant. There are mainly two categories of control: (i) airflow control and (ii) rotational speed control.

S. K. Mishra · M. R. Kumar (✉) · B. Appasani · A. V. Jha
School of Electronics Engineering, Kalinga Institute of Industrial Technology, Bhubaneswar, Odisha, India
e-mail: mano.kumarfet@kiit.ac.in

A. Pati
Department of Electrical Engineering, National Institute of Technology, Silchar, Assam, India

© The Author(s), under exclusive license to Springer Nature Switzerland AG 2023
O. Castillo and A. Kumar (eds.), *Recent Trends on Type-2 Fuzzy Logic Systems: Theory, Methodology and Applications*, Studies in Fuzziness and Soft Computing 425,
https://doi.org/10.1007/978-3-031-26332-3_7

The assessment of the state-of-the-art OWC technology, which explored a broad viewpoint from sea waves to grid connection, was presented in Delmonte et al. [1]. The Wells turbine and doubly fed induction generator (DFIG) used at the OWC wave power plant was described in Mishra et al. [2] as having improved efficiency due to a rotational velocity regulating mechanism designed to maximize turbine output power. Fuzzy Maximum Power Point Tracking (MPPT)-Backstepping Controller was the name of the scheme. The article [3] discussed studies on wave energy cost reduction strategies. The development and testing of dependable and controlled power take-off (PTO) technology for OWC converters was the main objective. For OWC wave PTO systems, nonlinear control algorithms were put out in Mishra et al. [4]. Wells turbine, DFIG, and irregular wave models made up the system. A fuzzy logic controller (FLC) was suggested in Mishra et al. [5] for the OWC wave energy facility to stop the Wells turbine from stalling.

In Rajapakse et al. [6], it was suggested that a battery bank may be connected directly to the dc-linkage of a power conversion scheme, which would smooth the energy sent to the grid. Model predictive controllers (MPCs) for the rectification and conversion of the power converter were also developed to keep the rotor velocity inside its optimal range. In Mishra et al. [7], a Wells turbine model was produced using MATLAB's Xilinx System Generator (XSG) package. Lekube et al. [8] worked on a distinct MPPT control approach for the OWC in order to maximize the quantity of electricity transmitted to the grid. Another study [9] discussed the pattern, modelling, and regulation of the OWC in order to maximize the NEREIDA power plant's energy output. A sliding mode control (SMC) strategy utilizing DFIGs, and Wells turbines was developed in Barambones et al. [10]. Contributions included: (i) an adaptive SMC structure that eliminated the need to calculate the system uncertainty bounds and (ii) a Lyapunov study of stability for the control method.

An optimization based MPPT approach for choosing the proper external rotor resistances of wound rotor induction generators (WRIG) was developed in Mishra et al. [11]. The study in Mishra et al. [12] sought to quantitatively assess the power production of an OWC placed on the south Brazil coast. Mishra et al. [13] investigated event-triggered backstepping controllers (ET-BSC) and event-triggered sliding mode controllers to control the rotating speed (ET-SMC). The study [14] concerned with the optimization of a wave energy converter's (WEC) efficiency. The WEC was an OWC device, and it was powered by a bidirectional flow impulse-turbine connected to a permanent magnet synchronous generator (PMSG). A lower order model of an OWC was reported in Suchithra et al. [15]. The strategy involved modelling of the capture chamber's hydrodynamic and aerodynamic coupling.

A centralized airflow management technique for a complex ocean energy system is detailed in Mishra et al. [16] in order to reduce the output power variance. The paper offers a straightforward method in Mishra et al. [17] for creating a PID controller to manage the OWC system's turbine velocity. The PID controller's goal was to maximize the amount of wave energy extracted from OWC. A unique Fuzzy Gain Scheduled- SMC was used in M'Zoughi et al. [18] to simulate an OWC and a rotating velocity control for stall-free operation. A FLC and airflow reference generator were constructed as well as verified in a simulation environment in Napole et al.

[19] to show how an OWC system's efficiency may be increased by controlling the turbine speed. The nonlinear model predictive controller developed in Magana et al. [20] maintains the efficacy of a self-rectifying turbine coupled to an OWC while maximizing the electro-mechanical power generated. An array of OWC WECs and nonlinear dynamics were modelled in state-space in Gaebele et al. [21]. To keep a reference turbine angular speed, the second-order SMC, which produces a smooth torque signal to a generator, was designed. The control mechanisms for a small WEC based on a PMSG were the main topic of Noman et al. [22]. It assesses the effectiveness of two control strategies: robust adaptive control (RAC) and conventional field-oriented control (FOC). In Roh and Kim [23], deep learning methods were utilized in the study to forecast the turbine generator's rotational speed in an OWC-WEC.

In the above-mentioned studies, airflow control as well as rotational speed control have been implemented. Linear as well as nonlinear control techniques were proposed. The Type 1 FLC was studied in Mishra et al. [5], M'Zoughi [18], and Napole et al. [19]. To the best of authors' knowledge, Type 2 FLC (T2FLC) is yet to be explored. Therefore, this research designs a T2FLC for OWC ocean energy plants for extracting the maximum energy. The controller works as a feedforward rotational speed controller wherein rotor speed tracks the reference speed. The reference rotor speed is generated using an MPPT algorithm that takes air velocity as input and provides reference speed as output. Finally, the realistic Joint North Sea Wave Project (JONSWAP) ocean wave model has been taken into account for the thorough simulations. The suggested T2FLC has been used to examine important OWC plant metrics such turbine power, generated power, turbine flow coefficient, and rotor speed. The rest of the chapter is organized as follows: Sect. 2 discusses the modeling overview of OWC plant; Sect. 3 describes the MPPT algorithm and design of the T2FLC; Simulation results are presented in Sect. 4 whereas conclusion is given in Sect. 5.

2 Modeling of OWC Plant

As seen in Fig. 1 [17], the OWC system is often built on the ocean's edge. The OWC is surrounded by four walls. Where the waves of the ocean crash into it, it is uncovered. The head of the column is filled with air, and the OWC tank is partially filled. A DFIG attached to the Wells turbine drives it, which is housed in a cylindrical container at the top of the chamber. The chamber's air is squeezed and released as a result of changes in seawater levels. As a result, the bidirectional airflow is brought on by the water's oscillation. The Wells turbine is designed in such a way that its spinning is always one way, despite the airflow's alternate course. Then, descriptions of the mathematical equations for the OWC tank, DFIG, Wells turbine, and ocean waves follow.

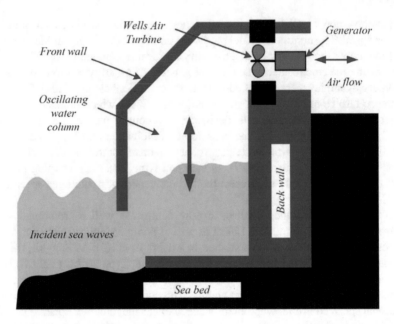

Fig. 1 Schematic representation of OWC plant

2.1 Wave Model and Chamber Dynamics

The JONSWAP model of the sea waves, one of the simulations often used in research, was employed to assess the efficacy of the suggested control systems [17]. Figure 2 depicts the JONSWAP wave band, which has a peak frequency of about 0.5 rad/s.

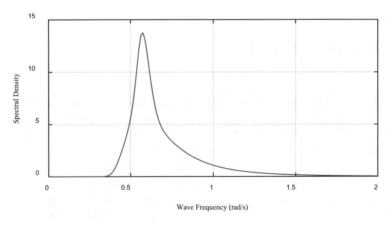

Fig. 2 JONSWAP wave spectrum

The OWC contains a sealed chamber with top and bottom apertures, as well as four side walls (Fig. 1). The container's bottom, which is half buried in water, gets struck by the waves. A two-way air flow is caused by the rising and decrease of the sea level. The wave height is the measurement of this difference in sea depth. The mathematical expression for air velocity is given by Mishra et al. [17]:

$$V_x = \left(\frac{A_{OWC}}{A_{duct}} \right) . \frac{\partial h(t)}{\partial t} \tag{1}$$

The representation for air velocity as presented in (1), provides the input to Wells turbine.

2.2 Wells Turbine Dynamics

Self-rectifying turbines include the Wells turbine [17]. Air flow supply is bidirectional, yet it still goes in the same way. Its stalling behavior is a drawback, though. The torque of the turbine is presented as Napole et al. [19]:

$$T_t = f_t(\phi) . V_x^2 \tag{2}$$

where, $f_t(\phi)$, a function of turbine flow coefficient, ϕ, can be stated as:

$$f_t(\phi) = C_t . k_t . r . (1 + \phi^{-1}) \tag{3}$$

C_t characterizes the Wells turbine characteristics. The C_t varies with ϕ as shown in Fig. 3. The turbine flow coefficient, ϕ, is provided by:

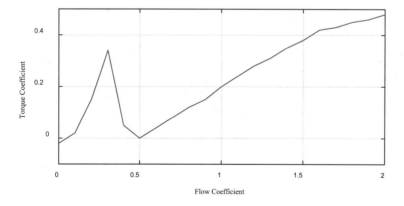

Fig. 3 Turbine characteristics

$$\phi = V_x.(r\omega_r)^{-1} \tag{4}$$

As demonstrated in Fig. 3, $\phi \leq \phi_{th} = 0.3$ offers highest torque coefficient, which would therefore deliver the most turbine torque and hence the greatest production power.

The Wells turbine is attached to a DFIG. So, the turbo-generator equation is given by:

$$\frac{d\omega_r}{dt} = \frac{1}{J}(T_t - F \cdot \omega_r - T_e) \tag{5}$$

where, ω_r is rotor speed; F is frictional coefficient; T_e is the electro-magnetic torque of DFIG.

2.3 DFIG Dynamics

A dynamic direct-quadrature (dq) DFIG version was taken into account in this investigation. The benefit of the dq model is that all three phases may be represented as dc quantities inside a stationary framework using a synchronous rotating frame [16]. The state equations of DFIG are given as:

$$\frac{d\psi_{ds}}{dt} = -\frac{R_s L_r}{K}\psi_{ds} + \omega_e\psi_{qs} + \frac{R_s L_m}{K}\psi_{dr} + v_{ds} \tag{6}$$

$$\frac{d\psi_{qs}}{dt} = -\omega_e\psi_{ds} - \frac{R_s L_r}{K}\psi_{qs} - \frac{R_s L_m}{K}\psi_{qr} + v_{qs} \tag{7}$$

$$\frac{d\psi_{dr}}{dt} = \frac{R_r L_m}{K}\psi_{ds} - \frac{R_r L_s}{K}\psi_{dr} - (\omega_r - \omega_e)\psi_{qr} + v_{dr} \tag{8}$$

$$\frac{d\psi_{qr}}{dt} = \frac{R_r L_m}{K}\psi_{qs} + (\omega_r - \omega_e)\psi_{dr} + v_{qr} \tag{9}$$

where, $K = L_s L_r - L_m^2$. ψ_{ds}, ψ_{qs}, ψ_{dr} & ψ_{qr} are dq flux quantities. R_s & R_r are DFIG resistances whereas L_s, L_r & L_m are DFIG inductances. ω_e is the stator supply frequency, v_{ds}, v_{qs}, v_{dr} & v_{qr} are DFIG voltages. The flux states ψ_{ds}, ψ_{qs}, ψ_{dr} and ψ_{qr} have initial conditions ψ_{ds0}, ψ_{qs0}, ψ_{dr0} and ψ_{qr0} respectively. The electromagnetic torque and output power expressions are:

$$T_e = -M(\psi_{qs}\psi_{dr} - \psi_{ds}\psi_{qr}) \tag{10}$$

$$P_g = T_e\omega_r \tag{11}$$

where, $M = -\left(\frac{3}{2}\right)\left(\frac{p}{2}\right)\left(\frac{L_m}{K}\right)$. p is number of pole of DFIG.

3 Design of MPPT and T2FLC Algorithm

The Fig. 4 presents the complete schematic diagram of OWC plant including back-to-back converters at rotor side and its control scheme. To feed nearly about 30% of the power generated from the grid to the rotor of the DFIG, there are two AC/DC converters known as the grid side converter (GSC) and rotor side converter (RSC). This flow of power is controlled by rotational speed controller that is consists of MPPT and T2FLC. Next, the MPPT algorithm and T2FLC design details have been presented.

3.1 Design of MPPT Algorithm

The next algorithm is the best reference speed calculation algorithm [16]. The following processes are taken to compute it using information about air velocity:
Step 1: Calculate the value of V_x using (1).
Step 2: Calculate peak values of V_x as following:

$$\left.\begin{array}{l} if \quad V_x \neq 0 \ \text{and} \ \dot{V}_x = 0 \\ \quad V_{xp} = V_x \\ else \\ \quad V_{xp} = 0 \\ end \end{array}\right\} \tag{12}$$

Step 3: Pass V_{xp} through zero-order-hold (ZOH) as:

$$\overline{V}_{xp} = ZOH(V_{xp}) \tag{13}$$

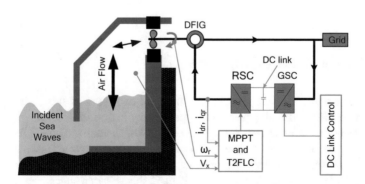

Fig. 4 MPPT and T2FLC scheme for OWC plant

Step 4: For threshold value of flow coefficient, $\phi_{th} = 0.3$, compute the reference speed using Eq. (4) as:

$$\omega_{ref} = \overline{V}_{xp}(r\phi_{th})^{-1} \tag{14}$$

Step 5: Calculate ω_{1d} by limiting the minimum and maximum values of ω_{ref} as:

$$\left.\begin{array}{l} if \quad \omega_{ref} \le \omega_e \\ \quad\quad \omega_{1d} = \omega_e \\ elseif \ \omega_{ref} \ge \omega_e \\ \quad\quad \omega_{1d} = \omega_{rp} \\ else \\ \quad\quad\quad \omega_{1d} = \omega_{ref} \\ end \end{array}\right\} \tag{15}$$

where, ω_e is the minimum value of ω_{1d} and ω_{rp} is the maximum value of ω_{1d}.

Step 6: To prevent any abrupt changes in reference value, pass ω_{1d} through a low pass filter. The new reference value is calculated as:

$$z_{1d} = h_f \otimes \omega_{1d} \tag{16}$$

The frequency domain representation of h_f is given as:

$$H_f(s) = \frac{1}{1 + 0.1s} \tag{17}$$

As a result, the z_{1d} is the optimum reference speed to be utilized for the proposed controllers. The next stage would be to create a backstepping type rotor speed regulator that would force the real rotor speed to follow the optimum reference speed z_{1d}.

3.2   Design of T2FLC Algorithm

Fuzzy sets are used in fuzzy logic, and there are two types of type 2 fuzzy logic that may be distinguished: generalized type-2 fuzzy logic and interval type-2 fuzzy logic. The computational complexity associated with generalized type-2 fuzzy logic is greatly reduced by interval type-2 fuzzy logic, which employs at least one interval type-2 fuzzy set. In engineering practice, interval type-2 fuzzy logic was therefore widely used. In order to implement T2FLC, this research suggests a zero-order interval type-2 fuzzy logic [24].

Figure 5 shows the graphical user interface of interval type 2 membership function. Here, input represents the rotor reference speed that is generated from the MPPT

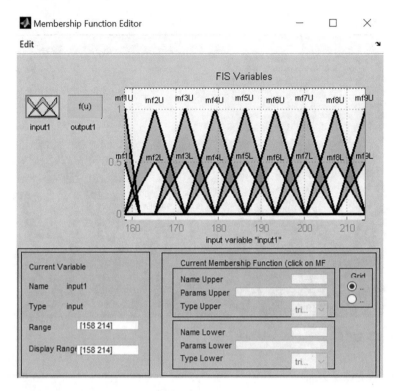

Fig. 5 T2FLC input membership function

block. The range of rotor speed is taken as [158 214] rad/s. The 9-membership function have been taken which are of triangular type with lower and upper bound. The 9 crisp type functions have been considered for output of T2FLC. Next, the inference engine rules are shown in Fig. 6 for the proposed system.

4   Results and Discussion

This section describes the mathematical simulations of T2FLC of OWC plant. The parameters for OWC plant simulation have taken from Mishra et al. [16]. First, the working of MPPT algorithm is discussed. Then, the comparison of T2FLC with uncontrolled plant is carried out. The JONSWAP irregular wave model has been considered that gives realistic sea scenario as shown in Fig. 7. Next, Fig. 8 displays the air velocity produced within the OWC chamber. The air velocity has unidirectional flow due to the self-rectifying nature of the Wells Turbine.

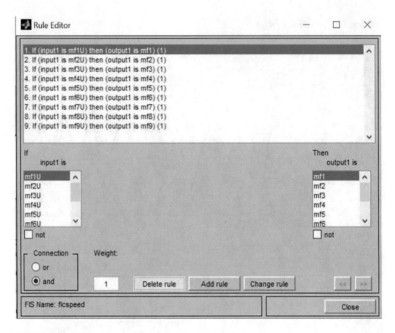

Fig. 6 T2FLC Takagi–Sugeno rule editor

Fig. 7 JONSWAP irregular wave

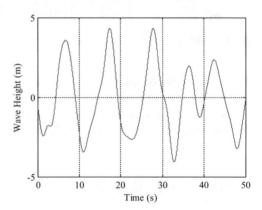

4.1 Performance of MPPT Algorithm

The MPPT algorithm is provided in previous section. First step is to generate air velocity waveform using JONSWAP irregular wave. Then, second step is to detect peak values of the air velocity followed by passing through the zero-order-hold (ZOH) in third step. This completes detection of peak values as shown in Fig. 9. The steps 4, 5 and 6 are used to evaluate the reference rotor speed which is further supplied to T2FLC block. The reference rotor speed is shown in Fig. 10.

Fig. 8 Air velocity inside
OWC chamber

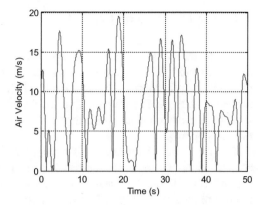

Fig. 9 Air velocity peak
values with ZOH

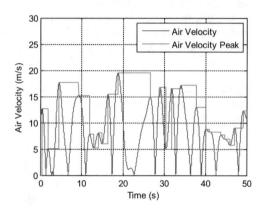

Fig. 10 Reference rotor
speed generated using MPPT
algorithm

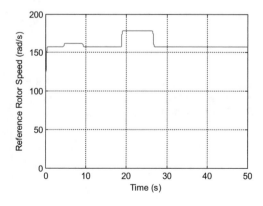

4.2 Performance of T2FLC

The performance of the T2FLC has now been examined and contrasted with that of the unregulated plant. Figure 11 shows the turbine flow coefficient for uncontrolled plant. On many instances, the value flow coefficient crosses the 0.3 limit which causes turbine stalling. As illustrated in Fig. 12, the T2FLC controlled OWC plant turbine flow coefficient remains within in the allowed limit, i.e. below 0.3 that prevents turbine stopping.

Further, in Fig. 13, the uncontrolled plant's rotor speed, which is caused by turbine stalling, hangs between 157 and 158 rad/s. The real rotor speed in the T2FLC-controlled OWC plant, as seen in Fig. 14, closely tracks the reference rotor speed to prevent turbine stalling and optimize output power. Figure 15 shows the control signal produced by the T2FLC.

The turbine torque for uncontrolled OWC plant is shown in Fig. 16. The peak turbine torque remains around 300 N-m for uncontrolled OWC plant whereas for T2FLC controlled OWC plant as shown in Fig. 17, it reaches upto 400 N-m.

Fig. 11 Turbine flow coefficient for uncontrolled OWC plant

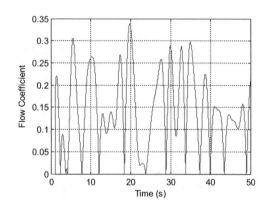

Fig. 12 Turbine flow coefficient for T2FLC controlled OWC plant

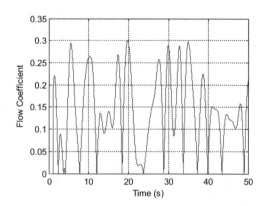

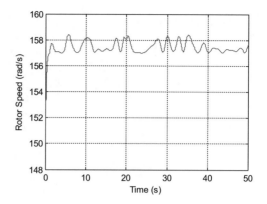

Fig. 13 Rotor speed for uncontrolled OWC plant

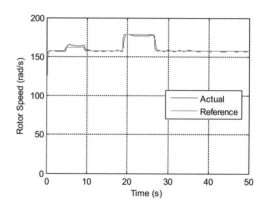

Fig. 14 Rotor speed for T2FLC controlled OWC plant

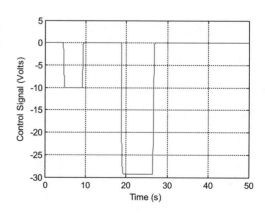

Fig. 15 Control signal for T2FLC controlled OWC plant

Finally, the output power waveforms generated for uncontrolled and T2FLC controlled OWC plant are shown in Figs. 18 and 19, respectively. The peak power

Fig. 16 Turbine torque for
uncontrolled OWC plant

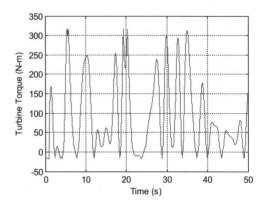

Fig. 17 Turbine torque for
T2FLC controlled OWC
plant

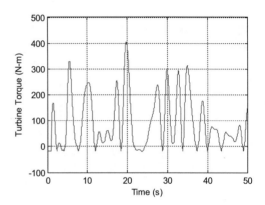

for uncontrolled OWC plant is around 50 kW whereas for T2FLC controlled OWC
plant, it is around 75 kW. This clearly shows the improvement of using proposed
control strategy.

Fig. 18 Output power for
uncontrolled OWC plant

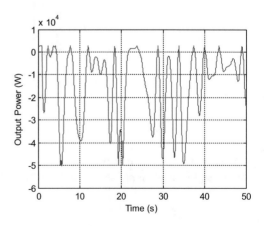

Fig. 19 Output power for
T2FLC controlled OWC
plant

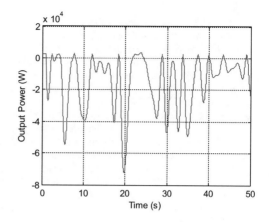

5 Conclusions

This book chapter suggested utilising T2FLC to manage the rotating speed of ocean energy plants using oscillating water columns. The T2FLC and MPPT algorithms provide the ocean waves' maximum electrical output power. The best rotor speed reference was thought to be produced using the MPPT algorithm. The reference speed was followed by the actual rotor speed thanks to the T2FLC. As a result, the turbine flow coefficient stays within the allowable range of 0.3, preventing the turbine from stalling. As a consequence, compared to an unmanaged OWC plant, the output power is maximized. Triangular membership functions of the interval type were used to achieve the T2FLC input fuzzification. For running simulations that closely resembled true sea scenarios, the JONSWAP irregular wave model was taken into consideration.

References

1. Delmonte, N., Barater, D., Giuliani, F., Cova, P., Buticchi, G.: Review of oscillating water column converters. IEEE Trans. Ind. Appl. **52**(2), 1698–1710 (2016). https://doi.org/10.1109/TIA.2015.2490629
2. Mishra, S.K., Purwar, S., Kishor, N.: An optimal and non-linear speed control of oscillating water column wave energy plant with wells turbine and DFIG. Int. J. Renew. Energy Res. **6**(3), 995–1006 (2016)
3. Garrido, I., Garrido, A.J., Lekube, J., Otaola, E., Carrascal, E.: Oscillating water column control and monitoring (2016). https://doi.org/10.1109/OCEANS.2016.7761420
4. Mishra, S.K., Purwar, S., Kishor, N.: Design of non-linear controller for ocean wave energy plant. Control Eng. Pract. **56**, 111–122 (2016). https://doi.org/10.1016/j.conengprac.2016.08.012
5. Mishra, S.K., Purwar, S., Kishor, N.: Fuzzy logic control of OWC wave energy plant for preventing wells turbine stalling (2017). https://doi.org/10.1109/POWERI.2016.8077219

6. Rajapakse, G., Jayasinghe, S., Fleming, A., Negnevitsky, M.: A model predictive control-based power converter system for oscillating water column wave energy converters. Energies **10**(10) (2017). https://doi.org/10.3390/en10101631

7. Mishra, S.K., Patel, A.: Wells turbine modeling and pi control scheme for OWC plant using Xilinx system generator. In: 2017 4th International Conference on Power, Control and Embedded Systems, ICPCES 2017, vol. 2017, Janua, pp. 1–6 (2017). https://doi.org/10.1109/ICPCES.2017.8117639

8. Lekube, J., Garrido, A.J., Garrido, I.: Rotational speed optimization in oscillating water column wave power plants based on maximum power point tracking. IEEE Trans. Autom. Sci. Eng. **14**(2), 681–691 (2017). https://doi.org/10.1109/TASE.2016.2596579

9. M'zoughi, F., Bouallègue, S., Garrido, A.J., Garrido, I., Ayadi, M.: Stalling-free control strategies for oscillating-water-column-based wave power generation plants. IEEE Trans. Energy Convers. **33**(1), 209–222 (2018). https://doi.org/10.1109/TEC.2017.2737657

10. Barambones, O., Gonzalez de Durana, J.M., Calvo, I.: Adaptive sliding mode control for a double fed induction generator used in an oscillating water column system. Energies **11**(11) (2018). https://doi.org/10.3390/en11112939

11. Mishra, S., Purwar, S., Kishor, N.: Maximizing output power in oscillating water column wave power plants: an optimization based MPPT algorithm. Technologies **6**(1), 15 (2018). https://doi.org/10.3390/technologies6010015

12. Lisboa, R.C., Teixeira, P.R.F., Torres, F.R., Didier, E.: Numerical evaluation of the power output of an oscillating water column wave energy converter installed in the southern Brazilian coast. Energy **162**, 1115–1124 (2018). https://doi.org/10.1016/j.energy.2018.08.079

13. Mishra, S.K., Purwar, S., Kishor, N.: Event-triggered nonlinear control of OWC ocean wave energy plant. IEEE Trans. Sustain. Energy **9**(4), 1750–1760 (2018). https://doi.org/10.1109/TSTE.2018.2811642

14. Suchithra, R., Ezhilsabareesh, K., Samad, A.: Optimization based higher order sliding mode controller for efficiency improvement of a wave energy converter. Energy **187** (2019). https://doi.org/10.1016/j.energy.2019.116111

15. Suchithra, R., Ezhilsabareesh, K., Samad, A.: Development of a reduced order wave to wire model of an OWC wave energy converter for control system analysis. Ocean Eng. **172**, 614–628 (2019). https://doi.org/10.1016/j.oceaneng.2018.12.013

16. Mishra, S.K., Appasani, B., Jha, A.V., Garrido, I., Garrido, A.J.: Centralized airflow control to reduce output power variation in a complex OWC ocean energy network. Complexity **2020** (2020). https://doi.org/10.1155/2020/2625301

17. Mishra, S.K., Appasani, B., Verma, V.K., Vidyakant Jha, A., Kumar, M.R., Pati, A.: PID control of the OWC plant to improve ocean wave energy capture (2020). https://doi.org/10.1109/PIICON49524.2020.9113007

18. M'Zoughi, F., Garrido, I., Garrido, A.J., De La Sen, M.: Fuzzy gain scheduled-sliding mode rotational speed control of an oscillating water column. IEEE Access **8**, 45853–45873 (2020). https://doi.org/10.1109/ACCESS.2020.2978147

19. Napole, C., et al.: Double fed induction generator control design based on a fuzzy logic controller for an oscillating water column system. Energies **14**(12) (2021). https://doi.org/10.3390/en14123499

20. Magana, M.E., Parlapanis, C., Gaebele, D.T., Sawodny, O.: Maximization of wave energy conversion into electricity using oscillating water columns and nonlinear model predictive control. IEEE Trans. Sustain. Energy (2021). https://doi.org/10.1109/TSTE.2021.3138159

21. Gaebele, D.T., Magana, M.E., Brekken, T.K.A., Henriques, J.C.C., Carrelhas, A.A.D., Gato, L.M.C.: Second order sliding mode control of oscillating water column wave energy converters for power improvement. IEEE Trans. Sustain. Energy **12**(2), 1151–1160 (2021). https://doi.org/10.1109/TSTE.2020.3035501

22. Noman, M., Li, G., Wang, K., Han, B.: Electrical control strategy for an ocean energy conversion system. Prot. Control Mod. Power Syst. **6**(1) (2021). https://doi.org/10.1186/s41601-021-00186-y

23. Roh, C., Kim, K.H.: Deep learning prediction for rotational speed of turbine in oscillating water column-type wave energy converter. Energies **15**(2) (2022). https://doi.org/10.3390/en1 5020572

24. Luo, G., Li, H., Ma, B., Wang, Y.: Design and experimental research of observer-based adaptive type-2 fuzzy steering control for automated vehicles with prescribed performance. Mechatronics **81** (2022). https://doi.org/10.1016/j.mechatronics.2021.102700

Intuitionistic Type-II Fuzzy Logic-Based Inference System and Its Realistic Applications to the Medical Field

Mukesh Kumar Sharma⊙ and **Nitesh Dhiman**⊙

Abstract In this chapter, we merge type-II fuzzy logic, and intuitionistic fuzzy logic in a broader way, to develop an intuitionistic type-II fuzzy logic. In intuitionistic type-II fuzzy logic, we deal the uncertainty, concerning with the truth values as well as false values. Intuitionistic type-II fuzzy logic is a stimulus of traditional intuitionistic type-I fuzzy logic in such a way that ambiguity is presented into linguistic variables that can be handled by using the truth grades and false grades with some hesitation margin. On the other hand, the proposed intuitionistic type-II fuzzy logic-based inference system accommodate intuitionistic fuzzy IF–THEN rules, which holds intuitionistic type-II fuzzy sets (IFT$_y$(II)). We also discussed the applications of proposed system in the various fields including; engineering, medical and agriculture etc. We applied over methodology over a data of lung cancer patients, which consists sixteen medical entities of infected patients. We also gave an example to illustrate our proposed technique.

Keywords Intuitionistic type-II fuzzy set ((IFTy(II)) · Intuitionistic type-II fuzzy inference system · Lung cancer · MATLAB

1 Introduction

The fuzzy logic was given by Zadeh [1] in 1965, as a universality of the traditional notion of a bi-valued set and a premise to shelter the degree of truthfulness, as demonstrated in human wording [2, 3]. Despite the extensive utilize of type-1 fuzzy set, earlier works have settled that type-I fuzzy logic models ambiguity to a fixed degree in real-life utilizations [4]. Zadeh [5] gave a generalization of his previously given type-I fuzzy set (T_y(I)) theory, to incorporate type-II fuzzy set (T_y(II)) theory accomplish of handling uncertainties where T_y(I) grapples due to membership functions of T_y(II) are ourselves fuzzy sets. In the T_y(I) based system the knowledge used to build certain rules, with uncertain antecedents and consequents, and these

M. K. Sharma (✉) · N. Dhiman
Department of Mathematics, Chaudhary Charan Singh University, Meerut 250004, India
e-mail: drmukeshsharma@gmail.com

© The Author(s), under exclusive license to Springer Nature Switzerland AG 2023
O. Castillo and A. Kumar (eds.), *Recent Trends on Type-2 Fuzzy Logic Systems: Theory, Methodology and Applications*, Studies in Fuzziness and Soft Computing 425,
https://doi.org/10.1007/978-3-031-26332-3_8

uncertainty present in this system translate into membership functions. Some other theories exist in which the uncertainty of the antecedent and consequent parts is in $T_y(II)$ form. $T_y(II)$ is more applicable in such situations in which, it is quite hard to find accurate membership values for a given fuzzy set. Later, Atanassov [6] gave the theory of intuitionistic fuzzy logic, to enhance Zadeh's fuzzy set theory, in which the uncertainty can be handle by taking the degrees of truthiness and falseness, with some hesitancy margin. Atanassov and Gargov [7] generalized the concept of intuitionistic fuzzy logic over interval valued intuitionistic fuzzy logic. Mendel et al. [8], used $T_y(II)$ based system whose degrees of truthiness are present in the form of intervals gives better performance in real-life applications in compare with the $T_y(I)$ based system. In the study of fuzzy set theory, $T_y(II)$ based logic is anextension of traditional fuzzy set-based logic i.e., $T_y(I)$ based logic. In $T_y(II)$ based logic, the uncertainty is not restricted over linguistic values, but also existing in the form of the truthinessvalues. On the contrary, intuitionistic fuzzy sets can also be examined as generalization of $T_y(I)$. To represent the uncertainty of belonging of element of a universal set, the intuitionistic fuzzy setsnot restricted over truthiness values, but also a falseness value.

Nguyen et al. [9] introduced interval $T_y(II)$ in the context of intuitionistic fuzzy sets during the process cmeans clustering s being used. Soto et al. [10] gave a novel approach for time series prediction with the help of ensembles of artificial neuro fuzzy inference system-basedmodels using interval $T_y(I)$ and $T_y(I)$ integrators. Lin et al. [11] introduced a TSK-type-based self-evolving compensatory in the context of interval type-2 fuzzy neural network. Tung et al. [12] also conducted a study on e-type-II fuzzy inference system: an evolving by using type-II neural fuzzy inference system. Abiyev and Kaynak [13] introduced the concept to identify and control of time-varying plants based on type-II fuzzy neural structure. Lin et al. [14] simplified interval type-II fuzzy neural networks. A hierarchical type-II fuzzy logic control [15] have also been discussed for autonomous mobile robot. John and Czarnecki [16] introduced the concept of type-II adaptive fuzzy inferencing system and aspects related to given methodology. Khanesar et al. [17] gave a noise reduction property of $T_y(II)$ based logic systems with the help of some new type-II membership functions. Juang and Tsao [18] gave a self-evolving algorithm with parameter learning and online structure, based on interval type-II fuzzy neural network. A method of defuzzification for type-II fuzzy numbers [19] and its applications have also been discussed. Castillo et al. [20] gave a differential evolution algorithm with type-II fuzzy logic-based system. Naderipour et al. [21] introduced a type-II fuzzy community detection model in large-scale social networks.

We discuss the generalized concepts of fuzzy set theory, $T_y(II)$ based logic systems and intuitionistic type-II fuzzy set ($IFT_y(II)$) based logic. We will give a novel way to represent the $IFT_y(II)$. $T_y(II)$ based logic consists IF–Then rules, and the uncertainty is handled by $T_y(II)$. We will also introduce $IFT_y(II)$ based inference system. Finally, we will describe briefly the utilizations and problem of intuitionistic fuzzy modelling $T_y(II)$ based logic.

The proposed work is divided into nine sections, in the second section we gave concept of some basic definitions. In the third section, we gave the various step of the proposed algorithm based on $IFT_y(II)$ based logic. In the fourth section, mathematical formulation is given, a block diagram of the proposed model is also given in this section. In the fifth section, we gave the architecture of the $IFT_y(II)$ based logic inference system. In the sixth section, we described the process of data collection, and we also characterized the including factors into three/four linguistic form. In the seventh section, intuitionistic fuzzy rule for the proposed inference system. In the eighth section, we contain numerical computation section. In the last and ninth section, we gave the conclusion of our proposed work.

2 Basic Preliminaries

2.1 Type-II Fuzzy Set ($T_y(II)$)

A $T_y(II)$ is a fuzzy set, whose truthiness value contains uncertainty, i.e., the membership value is a fuzzy set. A $T_y(II)$ is defined as:

$$T_y(II) = \{((x, \varrho)\mu_{T_y(II)}(x, \varrho)): x \in X \ \& \ \varrho \in I_x \subseteq [0, 1]\}$$

where, I_x: Primary membership value and $\mu_{T_y(II)}(x, \varrho)$ is the secondary membership value, also $0 \le \mu_{T_y(II)}(x, \varrho) \le 1$. Geometrical representation of $T_y(II)$ is given in Fig. 1.

Example 1 Let $X = \{2, 4, 5\}$ be an age set with the primary membership of the points of X respectively, $I_2 = \{0.7, 0.6, 1.0\}$, $I_4 = \{0.5, 0.7, 0.8\}$, and $I_5 = \{0.5, 0.5, 0.6\}$. The secondary membership function of the point 2 is:

$$\mu_{T_y(II)}(2, \varrho) = (0.1/0.7) + (0.7/0.6) + (0.6/1.0)$$

i.e., $\mu_{T_y(II)}(2, 0.7) = 0.1$ is the secondary membership grade of 2 with the primary membership grade 0.2.

Fig. 1 Geometrical representation of Type-II Fuzzy set

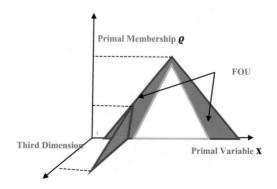

Similarly,

$$\mu_{T_y(II)}(4, \varrho) = (0.7/0.5) + (0.7/0.7) + (0.6/0.8)$$

$$\mu_{T_y(II)}(5, \varrho) = (0.8/0.5) + (0.6/0.5) + (0.6/0.6)$$

$T_y(II)$ can be expressed as:

$$T_y(II) = (0.1/0.7) + (0.7/0.6) + (0.6/1.0)/2 + (0.7/0.5) + (0.7/0.7)$$
$$+ (0.6/0.8)/4 + (0.8/0.5) + (0.6/0.5) + (0.6/0.6)/5.$$

2.2 Intuitionistic Type-II Fuzzy Set ($IFT_y(II)$)

A $IFT_y(II)$ is a set, whose membership and non-membership grades contains uncertainty, i.e., the membership function and non-membership grades are intuitionistic fuzzy sets. A $IFT_y(II)$ is defined as:

$$IFT_y(II) = \{((x, \varrho, \sigma), \mu_{IFT_y(II)}(x, \varrho), \nu_{IFT_y(II)}(x, \sigma)): x \in X \ \& \ \varrho \in I_x,$$
$$\sigma \in J_y \subseteq [0, 1]\}$$

where, J_x, is called the primary membership function, J_y is called the primal non-membership and $\mu_{T_y(II)}(x, \varrho)$ is the secondary membership function and $\nu_{IFT_y(II)}(x, \sigma)$, is called the secondary non-membership function also $0 \leq \mu_{IFT_y(II)}(x, \varrho) + \nu_{IFT_y(II)}(x, \sigma) \leq 1$.

Example 2 Let $X = \{2, 4, 5\}$ be an age set with the primary membership of the points of X respectively , $I_2 = \{0.7, 0.6, 1.0\}$, $I_4 = \{0.5, 0.7, 0.8\}$, and $I_5 = \{0.5, 0.5, 0.6\}$ & $J_2 = \{0.3, 0.2, 0\}, J_4 = \{0.5, 0.2, 0.1\}$, and $J_5 = \{0.5, 0.5, 0.2\}$. The secondary membership function of the point 2 is:

$$\mu_{IFT_y(II)}(2, \varrho) = (0.1/0.7) + (0.7/0.6) + (0.6/1.0);$$

$$\nu_{IFT_y(II)}(2, \varrho) = (0.7/0.3) + (0.2/0.2) + (0.4/0)$$

Similarly,

$$\mu_{IFT_y(II)}(4, \varrho) = (0.7/0.5) + (0.7/0.7) + (0.6/0.8);$$

$$\nu_{IFT_y(II)}(4, \varrho) = (0.2/0.5) + (0.2/0.2) + (0.3/0.1)$$

$\mu_{\text{IFT}_y(\text{II})}(5, \varrho) = (0.8/0.5) + (0.6/0.5) + (0.6/0.6);$

$$v_{\text{IFT}_y(\text{II})}(5, \varrho) = (0.1/0.5) + (0.2/0.5) + (0.2/0.2)$$

$\text{IFT}_y(\text{II})$ can be expressed as:

$$\text{IFT}_y(\text{II}) = \left\{ \left[\left(\frac{0.1}{0.7}\right) + \left(\frac{0.7}{0.6}\right) + \left(\frac{0.6}{0.1}\right) \right] 1/2, \left[\left(\frac{0.7}{0.3}\right) + \left(\frac{0.2}{0.2}\right) + \left(\frac{0.4}{0}\right) \right] 1/2 \right\}$$
$$+ \left\{ \left[\left(\frac{0.7}{0.5}\right) + \left(\frac{0.7}{0.7}\right) + \left(\frac{0.6}{0.8}\right) \right] 1/4, \left[\left(\frac{0.2}{0.5}\right) + \left(\frac{0.2}{0.2}\right) + \left(\frac{0.3}{1}\right) \right] 1/4 \right\}$$
$$+ \left\{ \left[\left(\frac{0.8}{0.5}\right) + \left(\frac{0.6}{0.5}\right) + \left(\frac{0.6}{0.6}\right) \right] 1/5, \left[\left(\frac{0.1}{0.5}\right) + \left(\frac{0.2}{0.5}\right) + \left(\frac{0.2}{0.2}\right) \right] 1/5 \right\}.$$

3 Proposed Algorithm for the Intuitionistic Type-II Fuzzy Logic Based Model

Step 1: Consider the set of input factors involve in the problem defined as:

$$I = \{I_1, I_2, \ldots\ldots I_n\}$$

Step 2: Applying fuzzification process to make triangular intuitionistic fuzzy number for each input factors given below:

$$IF = \{(\tau, \mu(\tau), v(\tau)) : \tau \in I; \mu, v : I \to [0, 1]\}$$

Step 3: Generate secondary membership and non-membership values denoted by $\mu_{\text{IFT}_y(\text{II})}(\tau, \varrho)$ and $v_{\text{IFT}_y(\text{II})}(\tau, \sigma)$ by giving a membership and non-membership grade to each input variable with set of primal membership value and primal non-membership value.

Step 4: After the step 3, we have the generalized $\text{IFT}_y(\text{II})$, $\text{IFT}_y(\text{II})$, is three-dimensional space defined as:

$$\text{IFT}_y(\text{II}) = \{((\tau, \varrho, \sigma), \mu_{\text{IFT}_y(\text{II})}(\tau, \varrho), v_{\text{IFT}_y(\text{II})}(\tau, \sigma)) : x \in I \& \varrho \in I_x,$$
$$\sigma \in J_y \subseteq [0, 1]\}$$

Also, consider the output factors '$\text{IFT}_y^{O_{ut}}(\text{II})$' (categories into m linguistic categories) in $\text{IFT}_y(\text{II})$ form, which denotes the final crisp output of the system.

Step 5: Now, for the inference system process, we need to use some rules based on knowledge-based system. For this process, consider the conditional and unqualified intuitionistic fuzzy proposition for obtained $\text{IFT}_y(\text{II})$ input factors in the form;

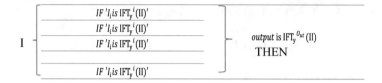

Step 6: The consequent part obtained from the intuitionistic fuzzy rule-based system, is present in $IFT_y(II)$ form. The reduces the $IFT_y(II)$ to convert it in the form of $IFT_y(I)$ by fuzzifying the area obtained in the third direction of the $IFT_y(II)$.

Step 7: Now, we obtained the output in the form of $IFT_y(I)$ using defuzzification process, we finally get the crips output of the proposed system.

4 Mathematical Formulation of the Problem

Consider a set of input variable 'τ' over a Universe of discourse X, formulate triangular membership function as well as for the input variables with the help of triangular $IFT_y(II)$ whose membership and non-membership grades are given as:

$$\mu_{tri}(\tau) = \begin{cases} \frac{\tau-x}{y-x} & x \leq \tau \leq y \\ \frac{z-\tau}{z-y} y & y \leq \tau \leq z \\ 0 & else \end{cases}$$

$$v_{tri}(\tau) = \begin{cases} \frac{\tau-\overline{x}}{y-\overline{x}} & \overline{x} \leq \tau \leq y \\ \frac{\overline{z}-\tau}{\overline{z}-y} y & y \leq \tau \leq \overline{z} \\ 0 & else \end{cases}$$

with $\overline{x} < x < y < z < \overline{z}$. Symbolically, to move from a $IFT_y(I)$ to an $IFT_y(II)$, a symbol is considered for the $IFT_y(II)$ put over the symbol for the intuitionistic fuzzy set; so, A denotes a $IFT_y(I)$, whereas $IFT_y(II)$ denotes the comparable $IFT_y(II)$. Zadeh [5] enhanced all of this to type-n fuzzy sets. The present work focuses only on $IFT_y(II)$ in order to develop a $IFT_y(II)$ based inference system. By giving a membership and non-membership grade to each input variable x with set of primal membership value ($\varrho \in I_x$) and primal non-membership value ($\sigma \in J_y$) to generate secondary membership and non-membership values denoted by $\mu_{IFT_y(II)}(x, \varrho) \& v_{IFT_y(II)}(x, \sigma)$. The membership and non-membership grades of a generalized $IFT_y(II)$, $IFT_y(II)$, is three-dimensional (considering secondary membership and non-membership values on the same ordinate), where the additional (third) dimension is the value of the secondary membership function and non-membership function at everyelement on its bi-dimensional domain known as its "footprint of uncertainty (FOU)". Now, develop

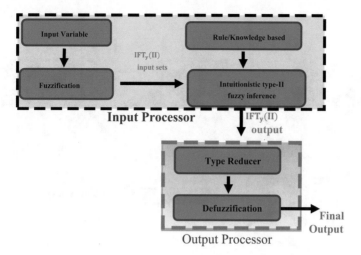

Fig. 2 Proposed intuitionistic type-II fuzzy inference system (block wise)

an $IFT_y(II)$ based inference system which consists two major components namely: (a) Input processor and (b) output processor, the input processor further consists following sub-components:

- The input variable
- Fuzzification and
- $IFT_y(II)$ based inference (based on $IFT_y(II)$ rules).

In the similar way the output processor further consists two sub-components as follows:

- $IFT_y(II)$ to $IFT_y(I)$ reducer and
- De-fuzzifier.

The geometrical representation of block diagram of the proposed system is given in Fig. 2.

5 Proposed Fuzzy Inference System

The architecture of the proposed $IFT_y(II)$ based inference system is given by Fig. 3.

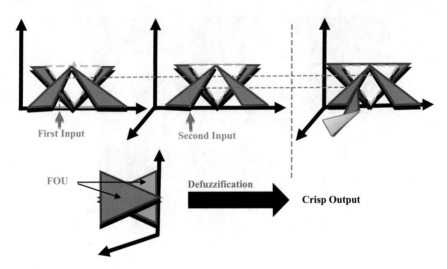

Fig. 3 Proposed intuitionistic type-ii fuzzy inference system

6 Data Collection Process

The secondary data of lung cancer infected patient is collected [22], out of 1000 lung cancer infected patient, we have taken 20 patients data set, to apply our approach to the conducted data set. The patients with the values of their various symptoms are given in Table 1.

Further, we have categorized the input and output factors in three and four linguistic categories. The categories are given in Table 2.

7 Intuitionistic Fuzzy Rules for the Proposed Intuitionistic Type-II Fuzzy Inference System

For the proposed intuitionistic type-II fuzzy system the conducted intuitionistic fuzzy rules, with sixteen input factors and one output factor is given by Table 3.

8 Numerical Computation

Consider the patient one with patient ID P1, with age 33, air pollution 2, alcohol use 4, dust allergy 5, genetic risk 3, balance diet 2, obesity 4, smoking level 3, chest pain 2, coughing of blood 4, fatigue 3, weight loss 4, shortness of breath 2, swallowing difficulty 2, dry cough 3, and coughing up.

Table 1 Collected data of lung cancer infected patients

Patient id	Age	Air pollution	Alcohol use	Dust allergy	Genetic risk	Balanced diet	Obesity	Smoking	Chest pain
P-1	33	2	4	5	3	2	4	3	2
P-10	17	3	1	5	4	2	2	2	2
P-100	35	4	5	6	5	6	7	2	4
P-1000	37	7	7	7	6	7	7	7	7
P-101	46	6	8	7	7	7	7	8	7
P-102	35	4	5	6	5	6	7	2	4
P-103	52	2	4	5	3	2	4	3	2
P-104	28	3	1	4	2	4	3	1	3
P105	35	4	5	6	6	5	5	6	6
P-106	46	2	3	4	4	3	3	2	4
P-107	44	6	7	7	7	7	7	7	7
P-108	64	6	8	7	7	7	7	7	7
P-109	39	4	5	6	5	6	6	6	6
P-11	34	6	7	7	6	7	7	6	7
P-110	27	3	1	4	3	3	3	2	4
P-111	73	5	6	6	6	6	5	8	5
P-112	17	3	1	5	4	2	2	2	2
P-113	34	6	7	7	6	7	7	7	7
P-114	36	6	7	7	7	6	7	7	7
Patient id	Age	Coughing of blood	Fatigue	Weight loss	Shortness of breath	Swallowing difficulty	Dry cough	Chronic lung disease	Level
P-1	33	4	3	4	2	3	3	2	L

(continued)

Table 1 (continued)

Patient id	Age	Air pollution	Alcohol use	Dust allergy	Genetic risk	Balanced diet	Obesity	Smoking	Chest pain
P-10	17	3	1	3	7	6	7	2	M
P-100	35	8	8	7	9	1	7	4	H
P-1000	37	8	4	2	3	4	7	7	H
P-101	46	9	3	2	4	4	2	6	H
P-102	35	8	8	7	9	1	7	4	H
P-103	52	4	3	4	2	3	3	2	L
P-104	28	1	3	2	2	2	4	3	L
P105	35	5	1	4	3	4	4	5	M
P-106	46	4	1	2	4	5	1	3	M
P-107	44	7	5	3	2	8	5	6	H
P-108	64	7	9	6	5	2	1	6	H
P-109	39	6	5	3	2	3	5	4	M
P-11	34	8	4	2	3	4	7	7	H
P-110	27	2	2	2	3	1	6	2	L
P-111	73	5	4	3	6	1	6	5	M
P-112	17	3	1	3	7	6	7	2	M
P-113	34	8	4	2	3	4	7	7	H
P-114	36	7	8	5	7	7	6	7	H

Table 2 Linguistic categorization of including factors

Age	For Membership value Young (Y):- [0–30] Medium (M):- [25–60] Old (O):- [>=55] For Non-membership value Young (Y):- [0–32] Medium (M):- [24–62] Old (O):- [>=52]	Chest Pain (CP)	For Membership value Lesser (Lr):- [0–3] Middle (Ml):- [2.5–6] Higher (Hr):- [5.5–10] For Non-membership value Lesser (Lr):- [0–3.2] Middle (Ml):- [2.2–6.2] Higher (Hr):- [5.2–10]
Air Pollution (AP)	For Membership value Lesser (Lr):- [0–3] Middle (Ml):- [2.5–6] Higher (Hr):- [5.5–10] For Non-membership value Lesser (Lr):- [0–3.2] Middle (Ml):- [2.2–6.2] Higher (Hr):- [5.2–10]	Coughing of Blood (CB)	For Membership value Lesser (Lr):- [0–3] Middle (Ml):- [2.5–6] Higher (Hr):- [5.5–10] For Non-membership value Lesser (Lr):- [0–3.2] Middle(Ml):- [2.2–6.2] Higher (Hr):- [5.2–10]
Alcohol Use (AU)	For Membership value Lesser (Lr):- [0–3] Middle (Ml):- [2.5–6] Higher (Hr):- [5.5–10] For Non-membership value Lesser (Lr):- [0–3.2] Middle (Ml):- [2.2–6.2] Higher (Hr):- [5.2–10]	Fatigue (FT)	For Membership value Lesser (Lr):- [0–3] Middle (Ml):- [2.5–6] Higher (Hr):- [5.5–10] For Non-membership value Lesser (Lr):- [0–3.2] Middle (Ml):- [2.2–6.2] Higher (Hr):- [5.2–10]
Dust Allergy (DA)	For Membership value Lesser (Lr):- [0–3]; Middle(Ml):- [2.5–6] Higher(Hr):- [5.5–10] For Non-membership value Lesser (Lr):- [0–3.2] Middle (Ml):- [2.2–6.2] Higher (Hr):- [5.2–10]	Weight Loss (WL)	For Membership value Lesser (Lr):- [0–3] Middle (Ml):- [2.5–6] Higher (Hr):- [5.5–10] For Non-membership value Lesser (Lr):- [0–3.2] Middle (Ml):- [2.2–6.2] Higher (Hr):- [5.2–10]
Genetic Risk (GR)	For Membership value Lesser (Lr):- [0–3] Middle (Ml):- [2.5–6] Higher (Hr):- [5.5–10] For Non-membership value Lesser (Lr):- [0–3.2] Middle (Ml):- [2.2–6.2] Higher (Hr):- [5.2–10]	Shortness of Breath (SB)	For Membership value Lesser (Lr):- [0–3] Middle (Ml):- [2.5–6] Higher (Hr):- [5.5–10] For Non-membership value Lesser (Lr):- [0–3.2] Middle (Ml):- [2.2–6.2] Higher (Hr):- [5.2–10]

(continued)

Table 2 (continued)

Balance Diet (BD)	For Membership value Bad (Y):- [0–3] Average (M):- [2.5–6] Good (O):- [5.5–10] For Non-membership value Bad (Y):- [0–3.2] Average (M):- [2.2–6.2] Good (O):- [5.2–10]	Swallowing Difficulty (SD)	For Membership value Lesser (Lr):- [0–3] Middle (Ml):- [2.5–6] Higher (Hr):- [5.5–10] For Non-membership value Lesser (Lr):- [0–3.2] Middle (Ml):- [2.2–6.2] Higher (Hr):- [5.2–10]
Obesity (OB)	For Membership value Lesser (Lr):- [0–3] Middle (Ml):- [2.5–6] Higher (Hr):- [5.5–10] For Non-membership value Lesser (Lr):- [0–3.2] Middle (Ml):- [2.2–6.2] Higher (Hr):- [5.2–10]	Dry Cough (DC)	For Membership value Lesser (Lr):- [0–3] Middle (Ml):- [2.5–6] Higher (Hr):- [5.5–10] For Non-membership value Lesser (Lr):- [0–3.2] Middle (Ml):- [2.2–6.2] Higher (Hr):- [5.2–10]
Smoking (SM)	For Membership value Lesser (Lr):- [0–3] Middle (Ml):- [2.5–6] Higher (Hr):- [5.5–10] For Non-membership value Lesser (Lr):- [0–3.2] Middle (Ml):- [2.2–6.2] Higher (Hr):- [5.2–10]	Chronic Lung disease (CL)	For Membership value Lesser (Lr):- [0–3] Middle (Ml):- [2.5–6] Higher (Hr):- [5.5–10] For Non-membership value Lesser (Lr):- [0–3.2] Middle (Ml):- [2.2–6.2] Higher (Hr):- [5.2–10]
Output		For Membership value For Non-membership value None (N):- [0–2.5]; None:- [0–2.7] Stage I:- [2.2–4.5]; Stage I:- [2.1–4.7] Stage II:- [4.2–6]; Stage II:- [4.1–6.2] Stage III:- >6; Stage III:- >6.2	

Table 3 Intuitionistic fuzzy rules for the proposed inference system

Fuzzy rules	Age	AP	AU	DA	GR	BD	OB	SM	CP
1	Y	Lr	Lr	Lr	Lr	G	Lr	N	Lr
2	Y	Lr	Lr	Lr	Lr	B	Ml	Lr	Lr
3	Ml	Lr	Lr	Ml	Ml	A	M	Lr	Lr
4	Ml	Lr	Lr	Hr	G	A	N	Hr	Lr
5	Ml	Ml	Lr	M	B	A	Lr	Lr	H
6	O	H	M	Hr	M	A	Hr	Lr	Lr
7	Y	Hr	M	Hr	Ml	A	Ml	Lr	Lr
8	O	Lr	Lr	Lr	Lr	A	Lr	N	Lr

(continued)

Table 3 (continued)

Fuzzy rules	Age	AP	AU	DA	GR	BD	OB	SM	CP
9	Ml	Lr	Lr	Lr	Lr	B	Lr	Lr	Lr
10	O	Lr	Hr	Hr	Hr	G	Ml	Lr	Hr
11	Y	Ml	Hr	Hr	Hr	B	N	Hr	Hr
12	Ml	Lr	Lr	Ml	B	A	Ml	Ml	Hr
13	O	Ml	Ml	Hr	Lr	A	Ml	Lr	Lr
14	Y	Lr	Hr	Hr	Hr	B	Hr	Hr	Hr
15	Y	Ml	Hr	Ml	Ml	A	Ml	Lr	Ml
16	Y	Ml	Lr	Hr	Ml	A	N	Ml	Ml
17	M	Lr	Hr	Hr	B	A	Hr	Hr	Hr
18	Ml	Ml	Ml	Ml	Lr	A	Ml	Ml	Ml
19	Y	Lr	Ml	Lr	Lr	G	Lr	N	Lr
20	O	Lr	Lr	Lr	Hr	A	Ml	Lr	Lr
Fuzzy rules	Age	CB	FT	WL	SB	SD	DC	CL	Output
1	Y	Lr	N	N	N	N	N	Lr	N
2	Y	M	Lr	Lr	Lr	Lr	Lr	Lr	Stage I
3	Ml	Lr	Lr	Lr	Ml	Ml	Ml	Ml	Stage I
4	Ml	N	N	Lr	N	N	Lr	Hr	Stage II
5	Ml	Lr	M	Lr	Lr	M	Lr	Ml	Stage II
6	O	Hr	Hr	Ml	Ml	Hr	Hr	Hr	Stage III
7	Y	Ml	Hr	Ml	Hr	Lr	Lr	Hr	Stage II
8	O	Lr	N	N	N	N	N	Lr	N
9	Ml	Ml	Ml	Ml	Lr	Ml	Lr	Lr	Stage I
10	O	Hr	Hr	Hr	Hr	Hr	Hr	Hr	Stage III
11	Y	N	N	Hr	N	Hr	Hr	Hr	Stage III
12	Ml	Ml	Ml	Ml	Ml	Ml	Hr	Ml	Stage I
13	O	Ml	Hr	Lr	Ml	Mr	Hr	Hr	Stage I
14	Y	Hr	Hr	Hr	Hr	Hr	Hr	Hr	Stage III
15	Y	Ml	Ml	Hr	Hr	Lr	Ml	Ml	Stage II
16	Y	Ml	Ml	Ml	N	Ml	Ml	Hr	Stage I
17	M	Hr	Ml	Ml	Ml	Hr	Hr	Hr	Stage III
18	Ml	Ml	Ml	Ml	Ml	Ml	Ml	Ml	Stage I
19	Y	M;	N	M;	Lr	Lr	N	Lr	N
20	O	Lr	Hr	Lr	Hr	Lr	Lr	Lr	N

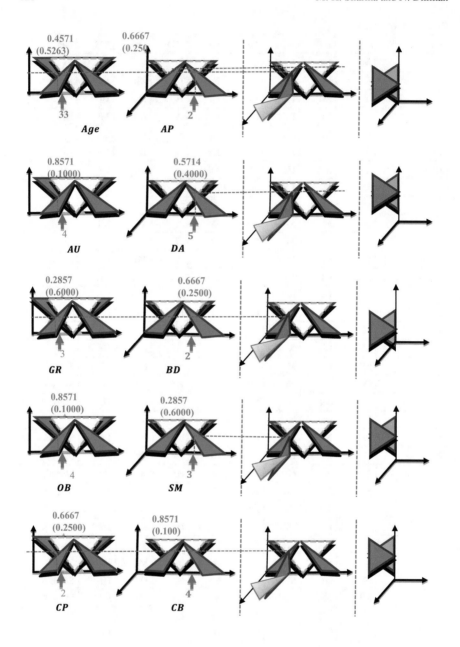

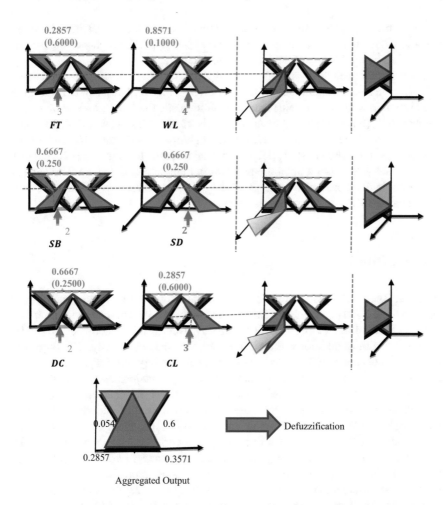

Aggregated Output

Let I = {33, 2, 4, 5, 3, 2, 4, 3, 2, 4, 3, 4, 2, 2, 2, 3} be set of inputs with membership values {0.4571, 0.6667, 0.8571, 0.5714, 0.2857, 0.6667, 0.8571, 0.2857, 0.6667, 0.8571, 0.2857, 0.8471, 0.6667, 0.6667, 0.666, 0.2857} and non-membership values {0.5263, 0.2500, 0.1000, 0.4, 0.6, 0.2500, 0.1000, 0.6000 0.2500, 0.100, 0.2500, 0.2500, 0.2500, 0.600} with respect to these input the output of the inference system is {0.4571/0.5263, 0.5714/0.4000, 0.2857/0.6000, 0.2857/0.6000, 0.6667/0.2500, 0.2857/0.6000, 0.2857/0.6000, 0.6667/0.2500}. Take minimum set of value, we get {0.2857, 0.6000}. Corresponds to this output the set of membership and non-membership values assign for the third dimension is {0.3571/0.054} and corresponding value of FOU is given in the form of triangular membership function [0.2857, 0.3214, 0.3571] and triangular non-membership function [0.054, 0.327, 0.6] after the defuzzification process [23, 24], we reduce the intuitionistic type-II fuzzy set in the form of intuitionistic type-I fuzzy set with membership and non-membership values of the output 0.3214 and 0.327. Now, taking the average of membership and non-membership values, we get the final output of the system i.e., 0.3242. The value is slightly large to the range of none category of the output factor; hence the patient is suffering by the Stage-I of the lung cancer category.

In the intuitionistic fuzzy systems generally employ intuitionistic type-I fuzzy sets $(IFT_y(I))$, on behalf of ambiguity by numbers in the range [0, 1]. But an $IFT_y(II)$ give rise to truth as well as false values that are intuitionistic fuzzy sets, instead of a crisp number. $IFT_y(II)$ offer chance to model levels of vagueness which $IFT_y(I)$ struggles. This additional dimension gives more degrees of liberty for improved picture of vagueness compared to $(IFT_y(I))$. So, the major reason to adopt the $IFT_y(II)$, due to additional way to think upon the problem in more appropriate way. The numerical example illustrates the advantage of $IFT_y(II)$ based inference system over then traditional $IFT_y(I)$ based inference systems.

9 Conclusion

We extend the $T_y(I)$ logic, in order to obtain the intuitionistic fuzzy logic, and further, develop an $IFT_y(II)$ based fuzzy inference system. We generalized the concept of $IFT_y(II)$ in the form of $IFT_y(II)$ in novel way. Due to this generalized $IFT_y(II)$, we can get a third additional dimensional for the conducted study. The objective behind the concept of $IFT_y(II)$ is to handle the uncertainty into linguistic variables form that can be handled by using the membership functions as well as the non-membership functions. We applied proposed concept over secondary data lung cancer patient. In this work, we have presented the novel concept of underlying $IFT_y(II)$ based logic and We have discussed in some detail $IFT_y(II)$ theory, fuzzy reasoning and develop an $IFT_y(II)$ based inference system with the Mamdani approach. In this study, an $IFT_y(II)$ based approach to the lung cancer patient's prediction is presented. The $IFT_y(II)$ based logic can accommodate more imprecision thereby imprecise knowledge and modelling better than the type-I fuzzy or intuitionistic fuzzy approaches. The key point in this design of intuitionistic type-II fuzzy logic-based inference system is to model the level of each lung cancer infected patient.

For the future prospective, we will show that how $IFT_y(II)$ based logic can be applicable to identify and the solution process of many real-world complex problems including engineering and agriculture fields etc. In future, we intend to learn the parameters of the $IFT_y(II)$ using Gaussian membership function and some hybrid approach to evaluate the real-world datasets. We will apply this concept over the Sugeno's type fuzzy inference system.

Acknowledgements The second author of this work is thankful to the "university grants commission, India", for the economic support.

References

1. Zadeh, L.A.: Fuzzy sets. Information and control, **8**(3), 338–353 (1965)
2. Pranevicius, H., Kraujalis, T., Budnikas, G., Pilkauskas, V.: Fuzzy rule base generation using iscretization of membership functions and neural network. In: Information and Software Technologies, pp. 160–171. Springer (2014)
3. Chiu, S.: Extracting fuzzy rules from data for function approximation and pattern classification. In: Fuzzy Information Engineering: A Guided Tour of Applications. Wiley (1997)
4. Mendel, J.M.: Uncertain rule-based fuzzy logic system: introduction and new directions (2001)
5. Zadeh, L.A.: The concept of a linguistic variable and its application to approximate reasoning-I. Inf. Sci. **8**, 199–249 (1975)
6. Atanassov, K.T.: Intuitionistic fuzzy sets. Fuzzy Sets Syst. **20**(1), 87–96 (1986)
7. Atanassov, K. Gargov, G.: Interval valued intuitionistic fuzzy sets. Fuzzy Sets Syst. **31**(3), 343–349 (1989)
8. Mendel, J.M., John, R.I., Liu, F.: Interval type-2 fuzzy logic systems made simple. IEEE Trans. Fuzzy Syst. **14**(6), 808–821 (2006)
9. Nguyen, D.D., Ngo, L.T., Pham, L.T.: Interval type-2 fuzzy c-means clustering using intuitionistic fuzzy sets. In: IEEE Third World Congress on Information and Communication Technologies (WICT), pp. 299–304 (2013)
10. Soto, J., Melin, P., Castillo, O.: A new approach for time series prediction using ensembles of ANFIS models with interval type-2 and type-1 fuzzy integrators. In: IEEE Conference on Computational Intelligence for Financial Engineering & Economics (CIFEr), pp. 68–73 (2013)
11. Lin, Y.-Y., Chang, J.-Y., Lin, C.-T.: A tsk-type-based self-evolving compensatory interval type-2 fuzzy neural network (TSCIT2FNN) and its applications. IEEE Trans. Industr. Electron. **61**(1), 447–459 (2014)
12. Tung, S.W., Quek, C., Guan, C.: eT2FIS: an evolving type-2 neural fuzzy inference system. Inf. Sci. **220**, 124–148 (2013)
13. Abiyev, R.H., Kaynak, O.: Type 2 fuzzy neural structure for identification and control of time-varying plants. IEEE Trans. Indus. Electron. **57**(12), 4147–4159 (2010)
14. Lin, Y.-Y., Liao, S.-H., Chang, J.-Y., Lin, C.-T.: Simplified interval type-2 fuzzy neural networks. IEEE Trans. Neural Netw. Learn. Syst. **25**(5), 959–969 (2014)
15. Hagras, H.A.: A hierarchical type-2 fuzzy logic control architecture for autonomous mobile robots. IEEE Trans. Fuzzy Syst. **12**(4), 524–539 (2004)
16. John, R.I., Czarnecki, C.: A type 2 adaptive fuzzy inferencing system. IEEE Int. Conf. Syst. Man Cybern. **2**, 2068–2073 (1998)
17. Khanesar, M.A., Kayacan, E., Teshnehlab, M., Kaynak, O.: Analysis of the noise reduction property of type-2 fuzzy logic systems using a novel type-2 membership function. IEEE Trans. Syst. Man Cybern. B Cybern. **41**(5), 1395–1406 (2011)
18. Juang, C.-F., Tsao, Y.-W.: A self-evolving interval type-2 fuzzy neural network with online structure and parameter learning. IEEE Trans. Fuzzy Syst. **16**(6), 1411–1424 (2008)
19. Biswas, A., De, A.K.: A unified method of defuzzification for type-2 fuzzy numbers with its application to multi-objective decision making. Granul. Comput. **3**, 301–318 (2018)
20. Castillo, O., Ochoa, P., Soria, J.: Differential Evolution Algorithm with Type-2 Fuzzy Logic for Dynamic Parameter Adaptation with Application to Intelligent Control, 1st edn. Springer International Publishing (2021)
21. Naderipour, M., Fazel, Z.M.H., Bastani, S.: A type-2 fuzzy community detection model in large-scale social networks considering two-layer graphs. Eng. Appl. Artif. Intell. **90**, 103206 (2020)
22. Nivedita, A.S., Sharma M.K.: Fuzzy mathematical inference system and its application in the diagnosis of lung cancer. Int. J. Agric. Stat. Sci. **17**(2), 709–717 (2021)
23. Klir, G.J., Clair, U.S., Yuan, B.: Fuzzy Set Theory: Foundations and Applications. Prentice-Hall, Inc. (1997)
24. Zimmermann, H.J.: Fuzzy Set Theory—And Its Applications. Springer Science & Business Media (2011)

An Enhanced Internet of Things Enabled Type-2 Fuzzy Logic for Healthcare System Applications

Joseph Bamidele Awotunde⬡, Olaiya Folorunsho, Isah Olawale Mustapha, Olayinka Olufunmilayo Olusanya, Mulikat Bola Akanbi, and Kazeem Moses Abiodun⬡

Abstract Due to advancements in information and communication technology, the Internet of Things has gained popularity in a variety of academic fields. In IoT-based healthcare systems, numerous wearable sensors are employed to collect various data from patients. The healthcare system has been challenged by the increase in the number of people living with chronic and infectious diseases. There are several existing IoT-based healthcare systems and ontology-based methods to judiciously diagnose, and monitor patients with chronic diseases in real-time and for a very long term. This was done to drastically minimize the vast manual labor in healthcare monitoring and recommendation systems. The current monitoring and recommendation systems generally utilised Type-1 Fuzzy Logic (T1FL) or ontology that is unsuitable owing to uncertainty and inconsistency in the processing, and analysis of observed data. Due to the expansion of risk and unpredictable factors in chronic and infectious patients such as diabetes, heart attacks, and COVID-19, these healthcare systems cannot be utilized to collect thorough physiological data about patients. Furthermore, utilizing the current T1FL ontology-based method to extract

J. B. Awotunde (✉)
Department of Computer Science, University of Ilorin, Ilorin, Nigeria
e-mail: awotunde.jb@unilorin.edu.ng

O. Folorunsho
Department of Computer Science, Federal University, Oye-Ekiti, Nigeria
e-mail: olaiya.folorunsho@fuoye.edu.ng

I. O. Mustapha
Department of Computer Science, Alhikmah University, Ilorin, Nigeria
e-mail: salnet2002@alhikmah.edu.ng

O. O. Olusanya
Department of Computer Science, Tai Solarin University of Education, Ijagun, Nigeria

M. B. Akanbi
Kwara State Polytechnic, Ilorin, Nigeria

K. M. Abiodun
Department of Computer Science, Landmark University, Omu Aran, Nigeria
e-mail: moses.abiodun@lmu.edu.ng

© The Author(s), under exclusive license to Springer Nature Switzerland AG 2023
O. Castillo and A. Kumar (eds.), *Recent Trends on Type-2 Fuzzy Logic Systems: Theory, Methodology and Applications*, Studies in Fuzziness and Soft Computing 425,
https://doi.org/10.1007/978-3-031-26332-3_9

133

the ideal membership value of risk factors becomes challenging and problematic, resulting in unsatisfactory outcomes. Therefore, this chapter discusses the applicability of IoT-based enabled Type-2 Fuzzy Logic (T2FL) in the healthcare system, and the challenges and prospects of their applications were also reviewed. The chapter proposes an IoT-based enabled T2FL system for monitoring patients with diabetes by extracting the physiological factors from patients' bodies. The wearable sensors were used to capture the physiological factors of the patients, and the data capture was used for the monitoring of patients. The results from the experiment reveal that the model is very efficient and effective for diabetes patient monitoring, using patient risk factors.

Keywords Type-2 fuzzy logic · Internet of Things · Ontology fuzzy logic · Healthcare systems · Patients monitoring · Risk factors · Chronic and infectious diseases

1 Introduction

The Internet of Things (IoT) based smart environments comprise various small devices enabled with computational features, sensors, and actuators [1, 2]. Various patients' physiological data can be collected using these devices connected to the IoT-based systems through a wireless network, and useful information can be extracted from them, and processing for decision [3, 4]. The emergence of the IoT-based wearable monitoring systems in smart healthcare systems is receiving suitable attention in recent times, and becoming popular these days in healthcare sectors [5].

In recent time, the number of chronic and infectious diseases suffering by patients globally is alarming, and require long-term planning for medical monitoring, and treatment continuously [6]. The situation has become serious for any patient or elderly to visit the hospital and in most cases unable to see their physicians [7, 8]. The IoT-based systems have become an effect to reduce the visitation of physicians by patients and offload patients from the hospital [9]. The IoT-based enabled various wearable sensors to orient and effective technique records vital physiological signs of the patient like electromyography (EMG), blood pressure, electrocardiography (ECG), heart rate, temperature, and gyroscope among others [10]. Other important features like contextual data around patients like humidity or temperature can be captured using sensors like infrared, camera, or contact [11].

But, due to the heterogeneous nature of IoT-based sensors and devices, the captured data produce erroneous and inaccurate data that resulted in weakening the interpretation accuracy, and energy proficiency of the system [12]. And most existing techniques are not proficient in extracting precise physiological features from the patients' body and participation values for bodily parameters as many of these systems use conventional technologies and techniques like fuzzy logic, classic ontology, and risk score calculators [13, 14]. Various data fusion has been proposed in recent times combining multi-source data enabled with ontology or fuzzy logic

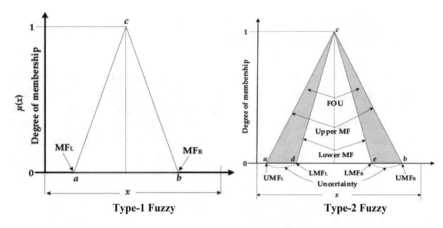

Fig. 1 The Type-1 fuzzy set and type-2 fuzzy set membership function [2]

for deciding to provide a better quality of service [15, 16]. However, these methods are inadequate due to their inconsistency and uncertainty of captured data, thus not efficient for accurate decision-making in healthcare systems. To sincerely deal with this issue, there should be the proper enumeration of the ambiguity in the sensor data (Fig. 1).

When considering the requirement of an IoT-based healthcare system for greater exemplification of patient information, imprecision and uncertainty must be eliminated. Existing methods based on ontology or type-1 fuzzy logic (T1FL) [17–20] for knowledge extraction from collected data with hazy data and uncertain characteristics are inefficient and ineffective. Because of its single membership value with the membership function, the T1FL cannot properly represent the uncertainty in the degree of membership. Because of its single membership function, the T1FL model is unreliable when there is a lot of uncertainty in the collected data, and fusing data becomes exceedingly difficult. The use of Type-2 fuzzy logic (T2FL) for decision-making in smart healthcare systems has recently been investigated. Because it is a simplified version of fuzzy logic, the T2FL may be used to correct large amounts of uncertainty in IoT-based source data. As a result, this chapter discusses the application of IoT-based enabled T2FL in healthcare systems, as well as the limitations and future possibilities of T2FL in dealing with the uncertainty inherent in decision-making systems' responses. This chapter's key contributions are as follows:

(i)   present the applicability of IoT-based enabled T2FL model in the healthcare system with their possible application challenges.
(ii) proposes an IoT-based enabled T2FL system for monitoring patients with diabetes by extracting the physiological factors from patients' bodies.
(iii) Prospects of IoT-based enabled T2FL model in smart healthcare systems.

2   Applications of Internet of Things and Type-2 Fuzzy Logic in Healthcare Systems

The rise of IoT-based systems as a new paradigm area of study has piqued the interest of researchers, academics, and government agencies [21]. The Internet of Things (IoT) includes an IP address for recognition, integrated into devices attached to a wireless network and connected to the Internet [22]. IoT-based devices are globally linked in real-time with other devices to gather data in real-time [23]. IoT-based technology implementation and applications are supported by many technologies such as embedded systems, Wireless Sensor Networks (WSNs) [24], cloud computing [25, 26], fog computing [27, 28], and edge computing [29, 30]. Their real-world applications include smart cities [31–33], smart transportation [34, 35], smart agriculture [36, 37], smart grid [38, 39], and smart supply chain management [40, 41]. The primary goal of IoT-based applications is to provide data in real-time for processing to make the right decisions and provide a sophisticated existence for humans [42].

The applications of IoT-based systems to real-world problems sometimes are characterized by high levels of numerical uncertainties and linguistics [43], and this has resulted in poor implementation of captured data using various devices attached to IoT-based systems. The emergence of fuzzy logic models (FLS) and their application to real-world applications have great success, especially in data with numerical uncertainties and linguistic [44]. The traditional T1FL model and ontology fuzzy logic have been the common fuzzy logic applied to various fields and been the vast majority of all fuzzy logic systems [45]. But, the T1FL cannot completely handle the high level of imprecision and uncertainties in IoT-based systems applications. This was so because of the use of precise and crisp fuzzy sets used by the T1FL model. Hence, the T2FL model was designed to handle a higher level of uncertainty for improved performance with a high level of accuracy [46, 47].

The T2FL's major goal and purpose are to manage uncertainty in information to increase the accuracy performance for the particular application it was designed for and of any system. The 2FL model surpasses the T1FL model in many aspects; nonetheless, their building principles are the same and follow practically the same approach [48]. The main variation between T1FL and T2FL models is the type of the membership functions [49]. The T2FL model employs stronger membership functions, which increase the model's effectiveness in dealing with the issue of uncertainty in real-world applications, particularly in IoT-based applications.

Once the membership function is selected, the fact that the genuine amount of membership is undetermined is no longer captured in the T1FL model [50]. Because the provided input is supposed to be in a precise condition with a single value of membership, uncertainty exists inside a specific application [43]. However, the dynamic unstructured environments linked with numerical and linguistic ambiguities generate challenges in forming rigorous and accurate experiences and consequent membership functions throughout the construction of a fuzzy logic system [43, 44]. Under particular operating settings and environments, the design of the T1FL model might become considerably sub-optimal, resulting in uncertainty and inaccurate findings in any application [51]. However, due to the associated uncertainties and changes in the environment, the selected T1Fl model sets may no longer be appropriate. This feature can lead to poor control and inefficiency due to degradation in T1FL model performance, resulting in the waste of resources spent on tuning the T1FL model or frequent redesigning of the model to find ways to deal with various uncertainties and issues that may arise when using the model [52, 53].

The conversion of T2FL into an equivalent T1FL fuzzy system was embraced by the output processor with an additional stage. This procedure was implemented using a Type-Reduction (TR) algorithm in the conversion of T2FL into T1FL fuzzy system. There are various arguments in favor of the T2FL system that show the potential abilities of the model to produce better application performance. The followings are the main reasons for the statement:

(i)   The T2FL model gives room for modification and increases the empathy of a rule-based system.
(ii)   The T2FL based on rules can help in dropping the complications involved in modeling a system.
(iii) The complex input/output relationship that seems impossible using the T1FL system can be obtained easily in the T2FL model.
(iv)   The T2FL system may be used to describe these input/output interactions without changing the number of rules involved in any way.

The T2FL system has been used and discovered pattern recognition, business and finance, electrical energy, risk indexing of power transformer breakdowns, and other fields are among them. real-world automatic control, modeling of micro-milling cutting forces, cyber security, and healthcare and medical. This chapter, on the other hand, focuses on healthcare and medical systems to cover many areas where the T2FL model is relevant for both short and long-term studies and applications.

The use of IoT-based enabled wearable monitoring systems in healthcare is a new technology, and a lot of academics have identified the problem of data fusion for IoT-based systems [54]. There are several wearable healthcare gadgets accessible for illness diagnostics, which are often smartphones and other devices [55]. The Internet of Medical Things (IoMT) refers to IoT-based medical equipment that connects with one another to distribute sensitive data [56]. The use of artificial intelligence has substantially aided in the processing of recorded data and transforming it into useful information that medical specialists may utilize for clinical decision-making and competence [57, 58]. The authors presented an IoT-based equipped with a comfortable method of wearability in [59].

The wearable sensors were utilized to gather physiological and environmental characteristics of the patient that may be used to monitor the patient's health state. The authors presented a method for analyzing and detecting heart attacks using fuzzy logic-based expert systems in [60]. When compared to other current models used in illness analysis and identification, the model outperforms them. The authors of

[61] presented an IoT-based real-time remote monitoring system for patients. The suggested solutions ensure the ECG's integrity in real-time, and a smartphone or computer may be used to access the monitor in real-time or by utilizing previously recorded data.

The authors proposed an IoT-based enabled wearable body sensor network for the diagnosis and monitoring of COVID-19 pandemic patients in Awotunde et al. [3]. The suggested technique is effective in monitoring COVID-19 patients by alerting medical experts and so making suitable decisions on what to do before the patient spreads the pandemic inside an environment. In Wang and Cai [62], the authors described a data networking-based IoT that was coupled with an edge cloud to create a secure healthcare monitoring system. The model used ciphertext and signature to increase the security of medical data transmission and hence the efficiency and effectiveness of data retrieval. System maintenance increases the security and privacy of the IoT-based environment while also ensuring that an attacker does not get access to the cloud. The authors of Reddy et al. [63] suggested a Firefly-BAT (FFBAT) and Artificial Neural Network (ANN) prediction model for feature selection using FFBAT and diabetes mellitus classification using the ANN model. The model was able to decrease the number of features by eliminating extraneous factors from the dataset, and ANN was used for dataset prediction. When compared to other existing models utilizing the same dataset, the suggested model performed better.

Another paper by the authors, Lee et al. [64], offered a T2FL-based smart diet recommendation. To recover meal records from a collection of recorded meals, the T2FL food ontology model was predefined. As a result, the ontology creation process was employed to carry out the acquisition of the T2FL personal food ontology. Each person's diet objectives were established, and T2FL assigns details to the various sorts of food. Then, to calculate the remaining calories for supper, the T2FL inference mechanism and fuzzy operators are combined with the T2FL personal food ontology. Finally, the personal food guide pyramid specified by field experts was utilized to develop the T2FL-based diet planning instrument, the T2FL-based menu recommendation machinery, and the T2FL-based semantic description mechanism to create a personal night meal plan. The suggested recommender outperformed the T1FL-based system and had promising performance when compared to existing ontology recommenders.

The authors of Habib et al. [65] presented a self-adaptive data gathering and fusion body sensor network for patient health monitoring. The biosensor was utilized to collect data, which was then sent to the fusion model for decision making using fuzzy set theory and a decision matrix. The suggested method enables medical specialists to make real-time and even distant decisions without physically viewing the patient. The authors of Muzammal et al. [66], Wu et al. [67] presented a multimodal device data synthesis approach to allow an ensemble technique from a body sensor network for medical data. The data collected from numerous sensors were merged and utilized to predict disease using ensemble classifiers. When compared to previous relevant works for disease prediction, the findings of the suggested system indicated a higher performance accuracy. The authors in Pinto et al. [68] suggested a genetic ML model enables parallel data in a wearable sensor network utilizing an IEEE 802.15.4 network. The suggested technique used the Genetic ML model to allow a trade-off between several user-defined metrics, which is the key benefit of the provided system. When compared to the Gur Game approach, the suggested strategy performs much better under different communication conditions and outperforms traditional periodic communication techniques over the IEEE 802.15.4 network. In Liu et al. [69], the author presented a DST-based fault-tolerant for event detection by looking at the impact of spatial correlation between nodes and their states on event detection. As each detector node's output was described, the proposed technique integrated them using weighted probability rather than crisp values, depending on their unique approach to identification. This allows the physician to avoid sending unneeded information by selecting and removing the affected parameters from the system.

3 The Challenges of the Internet of Things and Type-2 Fuzzy Logic in Healthcare Systems

One of the primary issues with IoT-based solutions in healthcare is data security and privacy [70]. The patients' impression of an IoT-based environment is critical to the platform's usability [71]. Most users feel that their data is not safe on the IoT-based platform; thus, utilizing it becomes tough for some of them since they believe that their information is not secure on the cloud. Identifying and addressing the security and privacy issues of IoT-based systems will go a long way toward altering the attitude of patients and other users, allowing them to have complete faith in the deployment of IoT-based healthcare systems. As a result, several studies, such as Awotunde et al. [72], Sajid et al. [73], Rizvi et al. [74], have been done in these areas.

The issue of data connection between the IoT-based systems and protocols during implementation needs to take with precaution and care. As the smart sensors integrated into IoT-based network increases, there is going to be a possible increase in the security hazard within the systems [75]. Though IoT increases the quality of community life and created business competitiveness globally, the IoT-based applications will surely attract attacks, hacking, and other possible cyber fraud. Research conducted by Hewlett Packard (2014) revealed over 70% of widely deployed IoT-based systems are significant defects with various forms of defaults. These flaws come in various ways like the disappointment of computer security, insecure network nodes, evasion in transport encryption, and authorization failure. Each computer on average contained at least 25 holes breaching the computer network connections. Hence, failure to protect the IoT-based applications in healthcare systems will bring serious opposition to the usage and acceptance of the system by individuals and medical sectors. IoT technology support critical strategic resources and utilities such as smart transportation, smart agriculture, and smart grid among others that need serious security services.

The IoT-based devices and sensors are not usually empowered with data encryption models that can protect the data transmission from one network node to another. The IoT-based systems produce a huge amount of data on patient wellbeing, household information, given financial details, and the situation of any organization, which can be processed and analyzed for decision-making within the organizations. Therefore, the security challenges can be reduced or eradicated by periodic training and educating the users on the need to take the security issues of the IoT-based systems with seriousness and the developers should always be educated to integrate security technologies like intrusion detection models, encryptions, and firewalls during software development. The training will teach users how to utilize the security features built into their smartphones and other connected devices.

When dealing with smart healthcare systems, the problem of privacy and protection must be thoroughly explored in order to prevent any compromises in client security and privacy. Security and privacy issues have arisen as a result of the massive volume of data created in the healthcare system, resulting in new risks and vulnerabilities in IoT-based platforms. When developing devices and sensors to be utilized by IoT-based applications, software developers should consider security mechanisms. The incorporation of security measures into computer software and hardware would undoubtedly aid in the reduction of security and privacy concerns in IoT-based systems. Designers' adoption of the client–server architecture will undoubtedly aid in data sharing in IoT-based contexts to ensure client data security and privacy. The usage of client–server architecture will aid in transmitting data and information to clients while keeping other sensitive data protected by a suitable certificate.

The applications T2FL model by various researchers in several fields have revealed positive results when compared with the traditional T1FL model. Despite these positive results of the T2FL technique, there are still some issues like the complexity of meet and join operation, and type-reduction occurrence limiting the appropriateness of the T2FL model in real-world applications. Generally, the processing times of T2FL are higher that T1FL systems due to the inference processing, and the computer overhead reduction. The interpretability of T2FL sets does not mask the type-reduction challenges due to the inherent complexity and redundancies in varied applications. The representation theorem for the T2FL model by Mendel and John [76] plays a significant role in describing T2FL set processes. But in practice, the theorem is more problematic in principle to apply and comprehend. At both theoretical and implementation levels, there is work currently ongoing on solving the processing time issue in the hardware and software stages. These works have been successful in reducing the T2FL model time issue, but there are still more works needed in the area of theoretical and algorithms implementation for the T2FL technique.

4 The Proposed Internet of Things-Based Enabled Type-2 Fuzzy Logic Framework

T2FL model for healthcare monitoring system presented based on IoT. The T2FL method calculates and visualizes the degree of uncertainty in various physiological data captured using sensor and devices. Combination concepts based on the T2FL system, the physiological properties of patients acquired using multiple devices and sensors are merged to enable the exact merging of heterogeneous sensor data. The suggested system was designed to effectively solve the problem of ambiguity and inconsistency of the T1FL model in the IoT-based healthcare system, the T2FL model was developed to handle various physiological parameters obtained from patients. Because the sensor data is inherently inconsistent and unreliable, the T1FL technique will not be effective for managing the acquired data utilizing such sensor devices. T2FL model is used to derive correct associative value from data acquired by multi-sensors on patients' physiological aspects. This was done to assess the data and offer an effective outcome that medical specialists might utilize for decision-making. Medical professionals may access the resulting patient data in real-time, remotely, and from any location for correct diagnosis and treatment.

The proposed IoT-based enabled T2FL system uses sensor-captured patient data as precise input to the T2FL model. Each sensor is analyzed by the five components of the T2LF model to determine the degree of membership: (i) fuzzifier, (ii) T2F rule, (iii) inference engine, (iv) reducer, and (v) defuzzifier. The T2FL rule applies the received crisp input data translated by fuzzifier to the "if–Then" rule. T2FL outputs were totally transformed to T1FL outputs using the type-reducer. Using defuzzification, the T1FL was further turned into a crisp value. The defuzzification results indicate the average points of the type-reduced set's right and left ends. Figure 2 depicts how fuzzification was utilized to convert sensor data into a T2FMFs interval and then into a T1Fs interval. The fuzzy membership values for the sensor data were calculated using the triangle membership function, with membership values of x, a, b, and c for every input.

$$f(x; a, b, c) = \begin{cases} 0, & x \leq a \\ \frac{x-a}{b-a} & a \leq x \leq b \\ \frac{x-c}{b-c} & b \leq x \leq c \\ 0, & c \leq x \end{cases} \tag{1}$$

$$f(x; a, b, c) = max\left(min\left(\frac{x-a}{b-a}, \frac{x-c}{b-c} \right), 0 \right) \tag{2}$$

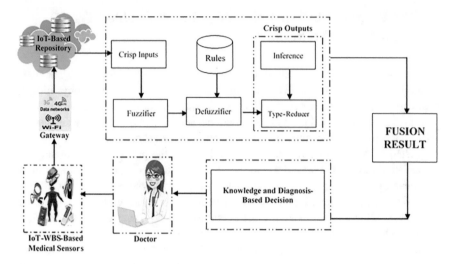

Fig. 2 The proposed IoT-based enabled type-2 fuzzy logic system

The Footprint of Uncertainty (FOU), Lower Membership Function (LMF), and Upper Membership Function (UMF) were described using four linear functions and five points a, b, c, d, and e. As a result, the following equations are used to express the centroid of the Type-2 fuzzy logic membership function.

$$traigular(x, a, b, c, d, e) = max(0, \ min(T_1, T_2, e)) \tag{3}$$

$$UMF = T_1(x, a, b, c) \tag{4}$$

$$LMF = T_1(x, d, e, c) \tag{5}$$

$$c = \frac{\sum_{i=1}^{q} \mu(x_i).x_1}{\sum_{i=1}^{q} \mu(x_i).} \tag{6}$$

$$\lfloor c_1, c_2 \rfloor = \left[\frac{\sum_{i=1}^{q} \mu'(x_i) \cdot x_1}{\sum_{i=1}^{q} \mu'(x_i)}, \ \frac{\sum_{i=1}^{q} \mu''(x_i) \cdot x_1}{\sum_{i=1}^{q} \mu''(x_i)} \right] \tag{7}$$

The $\mu'(x_i)$ and $\mu''(x_i)$ were used to signify the LMF and UMF values for the minimizes and maximizes weighted averages, respectively. Age (age), blood sugar (BS), weight (wgt), body temperature (BT), and patient health state are the raw inputs for diabetes prediction and monitoring (PHC). Because all of the variables are utilized for the input and are uncertain, the ranges of the values are set by the input variables for the border of the uncertainty of UMF and LMF. The T2FL model was used to compute the membership value of the collected data, and the results of the model's prediction provide the basis for the proposed system's knowledge engine to represent the membership value of each sensor data and process it for decision making. The T2FL model was utilized to deal with the uncertainty and inconsistency of sensor data from IoT-based systems, and the membership function output is transformed to mass value. The rules are fused to get the outcome for decision-making.

The finite set of mutually exclusive and exhaustive is denoted as $\Omega = \{\theta_1, \theta_2, \ldots, \theta_n\}$, Ω is the frame of discernment (FoD), 2^Ω represents the power set of Ω, and the 2^Ω elements can be formulated as.

$2^\Omega = \{\theta, \{\theta_1\}, \{\theta_2\}, \ldots, \{\theta_n\}, \{\theta_1, \theta_2\}, \ldots, \Omega\}$, the m is the mass function defined as a mapping from the power set 2^Ω to the interval of $[0, 1]$ fulfills the subsequent conditions:

$$\begin{cases} \sum_{A \leq \theta} m(A) = 1 \\ m(\theta) = 0 \end{cases} \tag{8}$$

A focal element happened if $mA > 1$, and mA is the mass function that characterizes how toughly the signal supports the propostion_A. the focal set and their associated mass value are represented by the Basic Probability Assignment (BPA) called BoE, or Basic Belief Assignment (BBA):

$$(\mathbb{R}, m) = \{A, \langle m \rangle : A \in 2^\Omega, m(A) > 0\}, \tag{9}$$

The subset of the set is $\mathbb{R}, 2^\Omega, A(\in \mathbb{R})$ each has an associated non-zero mass value $m(A)$.

The associate belief function, Bel can be used to denoted BPA, and plausibility function PI, and this can be formulated as follows:

$$Bel(A) = \sum_{\emptyset \neq B(\leq A)} m(B) \tag{10}$$

$$PI(A) = \sum_{B \cap A \neq \emptyset} m(B). \tag{11}$$

4.1 The Dataset Used for the Evaluation of the Proposed Model

The National Institute of Diabetes, Digestive and Kidney Diseases created the Pima Indians Diabetes Database dataset. The dataset was gathered to determine whether or not a patient has diabetes based on certain specified characteristics and diagnostic amounts in the dataset. The dataset's main limitation is that all of the patients are females who are at least 21 years old and have Pima Indian ancestry. There are many medical predictor characteristics and one outcome variable. These variables include

Table 1 Pima dataset
attribute characteristics

Name of the attribute	Categorized value
Age (year)	Age
Plasma Glucose Level	PGL
Diastolic Blood Pressure (mm Hg)	DBP
Body Mass Index (kg/m^2)	BMI
Frequency of pregnancy	Preg
Diabetes Pedigree Function	DPF
Thickness of Triceps Skin (mm)	TTS
2-h Serum Insulin	2HSI
Label class (0 or 1)	Class

BMI, Insulin level, Glucose, the number of pregnancies, and age, among others. As shown in Table 1, the dataset has 8 different features with 768 instances.

4.2 Performance Evaluation Metrics

The suggested model was assessed using four different performance metrics: accuracy, F1-score, sensitivity, and specificity. In statistics, the sensitivity for true positive rate and specificity for true negative rate define the performance of a binary classification test or classification function. The number of ill patients accurately diagnosed is measured by sensitivity, which is the percentage of sick patients that are correctly detected. The specificity is used to identify the number of patients who are accurately classed as healthy, or the proportion of healthy patients correctly detected. As a result, the metrics True Positive (TP), False Negative FP, True Negative (TN), and False Negative (FN) are computed. The following are the equations used for the computation of the metrics.

Accuracy is the ratio of correctly classified instances in percentage as presented by Eq. (12)

$$Accuracy = \frac{TP + TN}{TP + TN + FP + FN} \qquad (12)$$

The $F1_{Score}$ is the harmonic mean between the recall and precision as illustrated by Eq. (13)

$$F1_{Score} = \frac{2 \times Precision \times Recall}{Precision + Recall} \qquad (13)$$

Sensitivity is the proportion of the number of TP by the number of all of the positive estimations as shown by Eq. (14)

$$Sensitivity = \frac{TP}{TP + FN} \qquad (14)$$

Specificity is the proportion of the number of TN by the number of all of the positive estimations as shown by Eq. (15)

$$Specificity = \frac{TN}{TN + FP} \qquad (15)$$

Experimental Results

The suggested system was built using Python programming language on an Intel Core i7-2600 CPU with 8 GB RAM running Windows 10. The programming language collects input from diabetes datasets to determine the risk value of a diabetes patient. The PIMA diabetes datasets were classified using the Python computer language's Keras Library. The dataset was divided into several folds for training and testing using k-fold cross-validation techniques. The suggested model's efficacy was confirmed by comparing the findings to the performance of traditional fuzzy logic and T1FL-based models. Table 2 displays the suggested model's performance metrics findings utilizing four metrics, namely accuracy, F1-score, sensitivity, and specificity. Using the four measures, the suggested model greatly outperforms other models. Each folding of the dataset is used to determine the patient's diabetes status.

The results of the proposed system using cross-validation on the dataset give accuracy that is higher than 90% across all the folds, and the performance metrics used. The accuracy has the highest of 95.7% at tenfold, F1-score of 97.6 at tenfold, the sensitivity of 98.1% at nine-fold, and specificity of 97.1% at tenfold respectively. The proposed model performs best in terms of sensitivity of 98.1% but performs significantly better across all the metrics and the cross-validations.

Table 3 compares the proposed model to the T1FL scheme; the findings suggest that the T2FL model can reliably identify patients whether they have diabetes or not. The theory shows that the suggested model greatly outperforms the T1FL set across all performance criteria utilized to evaluate the models.

Table 2 The performance of the diabetes dataset using the four metrics

	Accuracy (%)	F1-score (%)	Sensitivity (%)	Specificity (%)
Twofold	93.3	93.6	95.5	93.0
Threefold	92.4	94.2	97.8	91.9
Fourfold	90.6	92.7	93.7	90.5
Fivefold	91.9	95.1	97.3	92.3
Sixfold	93.9	95.7	95.9	94.3
Sevenfold	91.4	92.1	93.6	92.1
Eightfold	92.6	94.8	96.3	92.9
Ninefold	92.3	95.3	98.1	91.6
Tenfold	95.7	97.6	97.4	97.1

Table 3 The comparison of T1FL with the proposed T2FL model

	T1FL scheme				T2FL scheme			
	Acc (%)	F1-S(%)	Sen. (%)	Spe. (%)	Acc. (%)	F1-S (%)	Sen. (%)	Spe. (%)
Twofold	98.3	86.8	88.7	89.2	93.3	93.6	95.5	93.0
Threefold	90.6	81.3	80.3	84.4	92.4	94.2	97.8	91.9
Fourfold	90.3	79.5	77.5	82.5	90.6	92.7	93.7	90.5
Fivefold	83.7	81.3	78.2	84.2	91.9	95.1	97.3	92.3
Sixfold	85.8	89.6	93.5	85.6	93.9	95.7	95.9	94.3
Sevenfold	89.9	87.6	84.7	89.7	91.4	92.1	93.6	92.1
Eightfold	89.4	86.9	84.3	89.3	92.6	94.8	96.3	92.9
Ninefold	83.7	81.6	79.3	84.2	92.3	95.3	98.1	91.6
Tenfold	87.2	84.6	83.6	87.8	95.7	97.6	97.4	97.1

5 Directions of Future Research

The applications of the T2FL model-enabled IoT-based system have witnessed a rise in recent times due to the acceleration in diagnosis, monitoring, treatment, and management of various diseases in healthcare systems. Though the huge amount of existing literature in these areas has enhanced the study of various applications of the T2FL model for diagnosis and recommendations, a lot still needs to be done in the areas of disease management and treatment using T2FL systems. This model can be used in drug management for various diseases, and treatments. There is still a need for further research to help in the implementation and enhancement of the approaches used for the T2FL model enabled IoT-based systems within healthcare systems. The potential research directions from the findings of this study are stated as follows.

The utilization and implementation of optimization in the area of T2FL are still a big open issue. The use of this will surely help in the application of T2FL in healthcare systems. The applications of choice of rules, defuzzification models, operations, type reduction, and membership functions are still carried out manually. There are still efficient opportunities of improving the T2FL model using evolutionary, learning algorithms, and metaheuristics models to enhance the T2FL within IoT-based applications in healthcare systems. The generation of big data from healthcare systems is no more a big issue, especially with the use of IoMT-based applications, the T2FL will very helpful in dealing with problems of uncertainty that will arise using various sensors in capturing various data. But this new trend will still need a lot of effort to deal with very critical issues in this area.

6　Conclusion

The issue of uncertainty and imprecision in the prediction and diagnosis of IoT-based based captured data using various sensors has been a challenge for many decades, and this has limited the huge benefits that would have been derived from this modern technology. The application of Fuzzy Logic model is one of the popular techniques use to improve the certainty, accuracy, and precision of IoT-based systems in healthcare sectors. However, T1FL model has not be able to deal with the problem of uncertainty that arise from IoT-based captured data from sensors, hence, T2FL sets have be prove useful in this direction. Therefore, this chapter review the applicability and issues of using IoT-based enabled T2FL scheme in the healthcare system, and future prospects of their applications in healthcare systems. The proposed an IoT-based enabled T2FL system for monitoring patients with diabetes using various sensors and devices to capture the physiological factors from patients' bodies. The results of the proposed model revealed that T2FL can be used to enhanced the problem of uncertainty that may arise from using captured data using sensors. The proposed model was evaluated using various performance metrics, and the results was compare with the T1FL model. The performance of the T2FL model was significantly better in term of accuracy, Sensitivity, specificity, and F1-score when compared with the T1FL model. The future work will look at the possibility of using optimization algorithms to enhance the performance of the T2FL model, the present work can still be extending in term of security, the IoT-based system can be secure using various security algorithms like encryption, blockchain, intrusion detection among others. This reduce the security concern by defend against all kinds of cybersecurity attacks like black hole, gray hole, denial of service and wormhole within the IoT-based platforms.

References

1. Awotunde, J.B., Ayoade, O.B., Ajamu, G.J., AbdulRaheem, M., Oladipo, I.D.: Internet of Things and cloud activity monitoring systems for elderly healthcare. Stud. Comput. Intell. **2022**(1011), 181–207 (2022)
2. Ullah, I., Youn, H.Y., Han, Y.H.: Integration of type-2 fuzzy logic and Dempster-Shafer theory for accurate inference of IoT-based healthcare system. Futur. Gener. Comput. Syst. **124**, 369–380 (2021)
3. Awotunde, J.B., Jimoh, R.G., AbdulRaheem, M., Oladipo, I.D., Folorunso, S.O., Ajamu, G.J.: IoT-based wearable body sensor network for COVID-19 pandemic. In: Advances in Data Science and Intelligent Data Communication Technologies for COVID-19, pp. 253–275 (2022)
4. Qiu, T., Chen, N., Li, K., Atiquzzaman, M., Zhao, W.: How can heterogeneous internet of things build our future: a survey. IEEE Commun. Surv. Tutor. **20**(3), 2011–2027 (2018)
5. Awotunde, J.B., Jimoh, R.G., Ogundokun, R.O., Misra, S., Abikoye, O.C.: Big data analytics of IoT-based cloud system framework: smart healthcare monitoring systems. Internet of Things **2022**, 181–208 (2022)
6. Wu, C.H., Lam, C.H., Xhafa, F., Tang, V., Ip, W.H.: IoT for Elderly, Aging and EHealth: Quality of Life and Independent Living for the Elderly, vol. 108. Springer Nature (2022)

7. Guo, X., Lin, H., Wu, Y., Peng, M.: A new data clustering strategy for enhancing mutual privacy in healthcare IoT systems. Futur. Gener. Comput. Syst. **113**, 407–417 (2020)
8. Uscher-Pines, L., Sousa, J., Raja, P., Mehrotra, A., Barnett, M.L., Huskamp, H.A.: Suddenly becoming a "virtual doctor": experiences of psychiatrists transitioning to telemedicine during the COVID-19 pandemic. Psychiatr. Serv. **71**(11), 1143–1150 (2020)
9. Aceto, G., Persico, V., Pescapé, A.: Industry 4.0 and health: Internet of things, big data, and cloud computing for healthcare 4.0. J. Indus. Inform. Integr. **18**, 100129 (2020)
10. Awotunde, J.B., Folorunso, S.O., Bhoi, A.K., Adebayo, P.O., Ijaz, M.F.: Disease diagnosis system for IoT-based wearable body sensors with machine learning algorithm. In: Hybrid Artificial Intelligence and IoT in Healthcare, pp. 201–222. Springer, Singapore (2021)
11. Ivanov, M., Markova, V., Ganchev, T.: An overview of network architectures and technology for wearable sensor-based health monitoring systems. In: 2020 International Conference on Biomedical Innovations and Applications (BIA), pp. 81–84. IEEE (2020)
12. Awotunde, J.B., Jimoh, R.G., Folorunso, S.O., Adeniyi, E.A., Abiodun, K.M., Banjo, O.O.: Privacy and security concerns in IoT-based healthcare systems. In: The Fusion of Internet of Things, Artificial Intelligence, and Cloud Computing in Health Care, pp. 105–134. Springer, Cham (2021)
13. Li, W., Chai, Y., Khan, F., Jan, S.R.U., Verma, S., Menon, V.G., Li, X.: A comprehensive survey on machine learning-based big data analytics for IoT-enabled smart healthcare systems. Mob. Netw. Appl. **26**(1), 234–252 (2021)
14. Chiang, T.C., Liang, W.H.: A context-aware interactive health care system based on ontology and fuzzy inference. J. Med. Syst. **39**(9), 1–25 (2015)
15. Du, J., Jing, H., Choo, K.K.R., Sugumaran, V., Castro-Lacouture, D.: An ontology and multi-agent-based decision support framework for prefabricated component supply chain. Inf. Syst. Front. **22**(6), 1467–1485 (2020)
16. Kalamkar, S., Geetha Mary, A.: Heterogeneous data fusion for healthcare monitoring: a survey. In: Big Data, IoT, and Machine Learning, pp. 205–232. CRC Press (2020)
17. Selvan, N.S., Vairavasundaram, S., Ravi, L.: Fuzzy ontology-based personalized recommendation for internet of medical things with linked open data. J. Intell. Fuzzy Syst. **36**(5), 4065–4075 (2019)
18. Collotta, M., Pau, G., Bobovich, A.V.: A fuzzy data fusion solution to enhance the QoS and the energy consumption in wireless sensor networks. In: Wireless Communications and Mobile Computing (2017)
19. Rasi, D., Deepa, S.N.: Energy optimization of Internet of Things in wireless sensor network models using type-2 fuzzy neural systems. Int. J. Commun. Syst. **34**(17), e4967 (2021)
20. Jana, D.K., Basu, S.: Novel Internet of Things (IoT) for controlling indoor temperature via Gaussian type-2 fuzzy logic. Int. J. Model. Simul. **41**(2), 92–100 (2021)
21. Ogundokun, R.O., Awotunde, J.B., Adeniyi, E.A., Misra, S.: Application of the Internet of Things (IoT) to fight the COVID-19 Pandemic. Internet of Things **2022**, 83–103 (2022)
22. Sennan, S., Ramasubbareddy, S., Balasubramaniyam, S., Nayyar, A., Abouhawwash, M., Hikal, N.A.: T2FL-PSO: Type-2 fuzzy logic-based particle swarm optimization algorithm used to maximize the lifetime of Internet of Things. IEEE Access **9**, 63966–63979 (2021)
23. Awotunde, J.B., Abiodun, K.M., Adeniyi, E.A., Folorunso, S.O., Jimoh, R.G.: (2021) A deep learning-based intrusion detection technique for a secured IoMT system. Commun. Comput. Inform. Sci. 1547 CCIS, 50–62
24. Adeniyi, E.A., Ogundokun, R.O., Awotunde, J.B.: IoMT-based wearable body sensors network healthcare monitoring system. Stud. Comput. Intell. **2021**(933), 103–121 (2021)
25. Awotunde, J.B., Bhoi, A.K., Barsocchi, P.: Hybrid cloud/fog environment for healthcare: an exploratory study, opportunities, challenges, and future prospects. In: Hybrid Artificial Intelligence and IoT in Healthcare, pp. 1–20. Springer, Singapore (2021)
26. Tang, J.: Discussion on health service system of mobile medical institutions based on Internet of Things and cloud computing. J. Healthc. Eng. (2022)
27. Alreshidi, E.J.: Introducing Fog Computing (FC) technology to Internet of Things (IoT) cloud-based anti-theft vehicles solutions. Int. J. Syst. Dyn. Appl. (IJSDA) **11**(3), 1–21 (2022)

28. Firouzi, F., Farahani, B., Marinšek, A.: The convergence and interplay of edge, fog, and cloud in the AI-driven Internet of Things (IoT). Inf. Syst. **107**, 101840 (2022)
29. Tang, Q., Xie, R., Yu, F.R., Chen, T., Zhang, R., Huang, T., Liu, Y.: Distributed task scheduling in serverless edge computing networks for the Internet of Things: a learning approach. IEEE Internet of Things J. (2022)
30. Ali, O., Ishak, M.K., Bhatti, M.K.L., Khan, I., Kim, K.I.: A comprehensive review of internet of things: technology stack, middlewares, and fog/edge computing interface. Sensors **22**(3), 995 (2022)
31. Malik, S., Gupta, D.: Examining the adoption and application of Internet of Things for smart cities. In: IoT and IoE Driven Smart Cities, pp. 97–119. Springer, Cham (2022)
32. Abiodun, M.K., Adeniyi, E.A., Awotunde, J.B., Bhoi, A.K., AbdulRaheem, M., Oladipo, I.D.: A framework for the actualization of green cloud-based design for smart cities. In: IoT and IoE Driven Smart Cities, pp. 163–182. Springer, Cham (2022)
33. Kamruzzaman, M.M., Alrashdi, I., Alqazzaz, A.: New opportunities, challenges, and applications of edge-AI for connected healthcare in internet of medical things for smart cities. J. Healthc. Eng. (2022)
34. Dogra, A.K., Kaur, J.: Moving towards smart transportation with machine learning and Internet of Things (IoT): a review. J. Smart Environ. Green Comput. **2**(1), 3–18 (2022)
35. Shamshuddin, K., Jayalaxmi, G.N.: Privacy-preserving scheme for smart transportation in 5G integrated IoT. In: ICT with Intelligent Applications, pp. 59–67. Springer, Singapore (2022)
36. Sinha, B.B., Dhanalakshmi, R.: Recent advancements and challenges of Internet of Things in smart agriculture: a survey. Futur. Gener. Comput. Syst. **126**, 169–184 (2022)
37. Rehman, A., Saba, T., Kashif, M., Fati, S.M., Bahaj, S.A., Choudhary, H.: A revisit of Internet of Things technologies for monitoring and control strategies in smart agriculture. Agronomy **12**(1), 127 (2022)
38. Dhaou, I.S.B., Kondoro, A., Kakakhel, S.R.U., Westerlund, T., Tenhunen, H.: Internet of Things technologies for smart grid. In: Research Anthology on Smart Grid and Microgrid Development, pp. 805–832. IGI Global (2022)
39. Krishnan, P.R., Jacob, J.: An IOT based efficient energy management in smart grid using DHOCSA technique. Sustain. Cities Soc. **79**, 103727 (2022)
40. Prajapati, D., Chan, F.T., Chelladurai, H., Lakshay, L., Pratap, S.: An Internet of Things embedded sustainable supply chain management of B2B e-commerce. Sustainability **14**(9), 5066 (2022)
41. Hrouga, M., Sbihi, A., Chavallard, M.: The potentials of combining Blockchain technology and Internet of Things for digital reverse supply chain: a case study. J. Clean. Prod. 130609 (2022)
42. Abikoye, O.C., Bajeh, A.O., Awotunde, J.B., Ameen, A.O., Mojeed, H.A., Abdulraheem, M., … & Salihu, S.A.: Application of internet of thing and cyber physical system in Industry 4.0 smart manufacturing. Adv. Sci. Technol. Innov. **2021**, pp. 203–217 (2021)
43. Hagras, H., Wagner, C.: Towards the wide spread use of type-2 fuzzy logic systems in real world applications. IEEE Comput. Intell. Mag. **7**(3), 14–24 (2012)
44. Hagras, H., Wagner, C.: Introduction to interval type-2 fuzzy logic controllers-towards better uncertainty handling in real world applications. IEEE Syst. Man Cybern. eNewsl. **27** (2009)
45. Dalpe, A.J., Thein, M.W.L., Renken, M.: PERFORM: a metric for evaluating autonomous system performance in marine testbed environments using interval type-2 fuzzy logic. Appl. Sci. **11**(24), 11940 (2021)
46. Mittal, K., Jain, A., Vaisla, K.S., Castillo, O., Kacprzyk, J.: A comprehensive review on type 2 fuzzy logic applications: past, present and future. Eng. Appl. Artif. Intell. **95**, 103916 (2020)
47. Melin, P., Castillo, O.: A review on type-2 fuzzy logic applications in clustering, classification and pattern recognition. Appl. Soft Comput. **21**, 568–577 (2014)
48. Karnik, N.N., Mendel, J.M.: Introduction to type-2 fuzzy logic systems. In: 1998 IEEE International Conference on Fuzzy Systems Proceedings. IEEE world congress on Computational Intelligence (Cat. No. 98CH36228), vol. 2, pp. 915–920. IEEE (1998)

49. Castillo, O., Melin, P., Kacprzyk, J., Pedrycz, W.: Type-2 fuzzy logic: theory and applications. In: 2007 IEEE International Conference on Granular Computing (GRC 2007), pp. 145–145). IEEE (2007)
50. Hagras, H.A.: A hierarchical type-2 fuzzy logic control architecture for autonomous mobile robots. IEEE Trans. Fuzzy Syst. **12**(4), 524–539 (2004)
51. Wijayasekara, D. S.: Improving understandability and uncertainty modeling of data using Fuzzy Logic Systems. Virginia Commonwealth University (2016)
52. Hagras, H.: Type-2 FLCs: a new generation of fuzzy controllers. IEEE Comput. Intell. Mag. **2**(1), 30–43 (2007)
53. Zhou, Y.S., Lai, L.Y.: Optimal design for fuzzy controllers by genetic algorithms. IEEE Trans. Ind. Appl. **36**(1), 93–97 (2000)
54. Folorunso, S.O., Awotunde, J.B., Ayo, F.E., Abdullah, K.K.A.: RADIoT: the unifying framework for IoT, radiomics and deep learning modeling. Intell. Syst. Ref. Libr. **2021**(209), 109–128 (2021)
55. Bajeh, A.O., Mojeed, H.A., Ameen, A.O., Abikoye, O.C., Salihu, S.A., Abdulraheem, M., ... & Awotunde, J.B.: Internet of robotic things: its domain, methodologies, and applications. Adv. Sci. Technol. Innov. **2021**, 135–146 (2021)
56. Papaioannou, M., Karageorgou, M., Mantas, G., Sucasas, V., Essop, I., Rodriguez, J., Lymberopoulos, D.: A survey on security threats and countermeasures in internet of medical things (IoMT). Trans. Emerg. Telecommun. Technol. e4049 (2020)
57. RM, S.P., Maddikunta, P.K.R., Parimala, M., Koppu, S., Gadekallu, T.R., Chowdhary, C.L., Alazab, M.: An effective feature engineering for DNN using hybrid PCA-GWO for intrusion detection in IoMT architecture. Comput. Commun. **160**, 139–149
58. Awotunde, J.B., Oluwabukonla, S., Chakraborty, C., Bhoi, A.K., Ajamu, G.J.: Application of artificial intelligence and big data for fighting COVID-19 pandemic. Decis. Sci. COVID-19, 3–26 (2022)
59. Haghi, M., Neubert, S., Geissler, A., Fleischer, H., Stoll, N., Stoll, R., Thurow, K.: A flexible and pervasive IoT-based healthcare platform for physiological and environmental parameters monitoring. IEEE Internet Things J. **7**(6), 5628–5647 (2020)
60. Muhammad, L.J., Algehyne, E.A.: Fuzzy based expert system for diagnosis of coronary artery disease in Nigeria. Heal. Technol. **11**(2), 319–329 (2021)
61. Yew, H.T., Ng, M.F., Ping, S.Z., Chung, S.K., Chekima, A., Dargham, J.A.: Iot based real-time remote patient monitoring system. In: 2020 16th IEEE International Colloquium On Signal Processing & Its Applications (CSPA), pp. 176–179. IEEE
62. Wang, X., Cai, S.: Secure healthcare monitoring framework integrating NDN-based IoT with edge cloud. Futur. Gener. Comput. Syst. **112**, 320–329 (2020)
63. Reddy, G.T., Khare, N.: Hybrid firefly-bat optimized fuzzy artificial neural network based classifier for diabetes diagnosis. Int. J. Intell. Eng. Syst. **10**(4), 18–27 (2017)
64. Lee, C.S., Wang, M.H., Hagras, H.: A type-2 fuzzy ontology and its application to personal diabetic-diet recommendation. IEEE Trans. Fuzzy Syst. **18**(2), 374–395 (2010)
65. Habib, C., Makhoul, A., Darazi, R., Salim, C.: Self-adaptive data collection and fusion for health monitoring based on body sensor networks. IEEE Trans. Industr. Inf. **12**(6), 2342–2352 (2016)
66. Muzammal, M., Talat, R., Sodhro, A.H., Pirbhulal, S.: A multi-sensor data fusion enabled ensemble approach for medical data from body sensor networks. Inform. Fus. **53**, 155–164 (2020)
67. Wu, T., Wu, F., Redoute, J.M., Yuce, M.R.: An autonomous wireless body area network implementation towards IoT connected healthcare applications. IEEE Access **5**, 11413–11422 (2017)
68. Pinto, A.R., Montez, C., Araújo, G., Vasques, F., Portugal, P.: An approach to implement data fusion techniques in wireless sensor networks using genetic machine learning algorithms. Inform. Fus. **15**, 90–101 (2014)
69. Liu, K., Yang, T., Ma, J., Cheng, Z.: Fault-tolerant event detection in wireless sensor networks using evidence theory. KSII Trans. Internet Inform. Syst. (TIIS) **9**(10), 3965–3982 (2015)

70. Awotunde, J.B., Chakraborty, C., Adeniyi, A.E.: Intrusion detection in industrial internet of things network-based on deep learning model with rule-based feature selection.Wirel. Commun. Mob. Comput. (2021)
71. Awotunde, J.B., Misra, S., Ayoade, O.B., Ogundokun, R.O., Abiodun, M.K.: Blockchain-based framework for secure medical information in Internet of Things system. In: Blockchain Applications in the Smart Era, pp. 147–169. Springer, Cham (2022)
72. Awotunde, J.B., Chakraborty, C., Folorunso, S.O.: A secured smart healthcare monitoring systems using blockchain technology. In: Intelligent Internet of Things for Healthcare and Industry, pp. 127–143. Springer, Cham (2022)
73. Sajid, A., Abbas, H., Saleem, K.: Cloud-assisted IoT-based SCADA systems security: a review of the state of the art and future challenges. IEEE Access 4, 1375–1384 (2016)
74. Rizvi, S., Orr, R.J., Cox, A., Ashokkumar, P., Rizvi, M.R.: Identifying the attack surface for IoT network. Internet of Things 9, 100162 (2020)
75. Awotunde, J.B., Misra, S.: Feature extraction and artificial intelligence-based intrusion detection model for a secure Internet of Things networks. In: Illumination of Artificial Intelligence in Cybersecurity and Forensics, pp. 21–44. Springer, Cham (2022)
76. Mendel, J.M., John, R.B.: Type-2 fuzzy sets made simple. IEEE Trans. Fuzzy Syst. 10(2), 117–127 (2002)

Artificial Neural Network Based Type-2 Fuzzy Optimization for Medical Diagnosis

Nitesh Dhiman, Nivedita, and Mukesh Kumar Sharma

Abstract In this work, we generalized the notion of fuzzy logic and neural network in order to develop type-2 neuro fuzzy system. With the help of proposed system, we will be able to deal medical problem to enhance the performance for reimbursing the higher uncertainties. For the validity of proposed system, we are giving numerical trials to justify our approach. Further, the parameters involve in the output of the proposed system is optimized by using teaching learning-based optimization (TLBO).

Keywords Type-2 fuzzy set · Neuro fuzzy system · Type-2 neuro fuzzy system · Teaching learning-based optimization (TLBO)

1 Introduction

In present era of vagueness Zadeh [1] gave the fuzzy logic approach to overcome the uncertainty. Professor Zadeh also gave the notion of a linguistic variable and its application [2] to estimated reasoning. Traditional neuro fuzzy systems have been functional in several real time applications. Due to the use of membership function only, it cannot be able to handle the uncertainly in a well manner. To remove such difficulty, we are adopting type-2 fuzzy set-based neuro fuzzy inference system. So, that the outcome can be more optimize with minimum error. The targeted output will be assumed to be 1 and the expected output can be found in such a manner that will give minimum error. The type-2 fuzzy set provide third dimension to study, and reduce the quantity of doubtful information in more appropriate way. The basic idea for this proposal is to get an error free or low error solution by the applicability of

N. Dhiman · M. K. Sharma (✉)
Chaudhary Charan Singh University, Meerut, UP 250004, India
e-mail: drmukeshsharma@gmail.com

Nivedita
Department of Mathematics, S.S.V. (PG) College, Hapur, India

Chaudhary Charan Singh University, Meerut, UP 250004, India

© The Author(s), under exclusive license to Springer Nature Switzerland AG 2023
O. Castillo and A. Kumar (eds.), *Recent Trends on Type-2 Fuzzy Logic Systems: Theory, Methodology and Applications*, Studies in Fuzziness and Soft Computing 425, https://doi.org/10.1007/978-3-031-26332-3_10

proposed type-2 neuro fuzzy system. To handle more uncertainty in a better mode, the generalization of type-1 fuzzy set and systems are considered in the form of type-2 fuzzy set. As we know, there is fixed truth function value in type-1 fuzzy set-based systems, whereas in type-2 fuzzy set, truth function value is fluctuating. Basically, a fuzzy set govern to input values so that they can be converted into fuzzy variables. To overcome the demerits of fuzzy sets, Zadeh in 1975, gave an idea of type-2 fuzzy set. The truth value of type-2 fuzzy set is 3 dimensional. In many circumstances, when we have doubt in deciding the truth grade as a number in [0, 1]. In these cases, type-2 fuzzy set provide needed framework to validate and work with this evidence. Xie and Lee [3] introduced an extended type-reduction method for general type-2 fuzzy sets. Castillo and Melin [4] gave the adaptive noise cancellation using type-2 fuzzy logic and neural networks. Castillo et al. [5] also provide an evolutionary computing for type-2 fuzzy systems to regulate the non-linear dynamic plants. Coupland and John [6] introduced a new and well-organized technique for the type-2 meet process in 2004. Hagras [7] gave hierarchical type-2 fuzzy logic control architecture for autonomous mobile robots. Hagras [8] also gave type-2 fuzzy logic system for self-directed mobile robots. Hwang and Rhee [9] introduced an interval valued type-2 fuzzy C spherical shells procedure. Figueroa et al. [10] have a type-2 fuzzy system to follow the mobile objects. Garibaldi et al. [11] provided a case study; the association between interval type-2 fuzzy sets and non-stationary. Lin et al. [12] introduced a type-2 fuzzy logic model to design buck DC-DC converters. Lynch et al. [13] have given an embedded type-2 FLC for real-time speediness control of marine and traction diesel machines. Later Mittal et al. [14] provided a review on type 2 fuzzy logic-based applications. Rhee and Hwang [15] introduced a type-2 fuzzy C-means clustering procedure. Rhee and Hwang [16] also gave an interval type-2 fuzzy perceptron with their applicability. Sepúlveda et al. [17] integrated a progress platform for brainy control based on type-2 fuzzy logic. Ontiveros-Robles et al. [18] gave a comparative analysis of noise robustness of type 2 fuzzy logic controllers. Ruiz-García et al. [19] introduced a concept on general forms of interval type-2 fuzzy sets. Ruiz-García et al. [19] introduced a concept on general forms of interval type-2 fuzzy sets. Sennan et al. [20] proposed a type-2 fuzzy logic-based particle swarm optimization algorithm based on Internet of Things.

Artificial intelligence (AI) has various branches like machine and deep learning; Knowledge based systems, fuzzy logics, robotics, and many more. Researchers want to choose their research area in the field of AI because of its various applications. In simplest form, artificial intelligence is when human behaviour is ascribed to intelligence. The applicability of type-2 fuzzy logic approach over various areas including science and technology plays a vital role in that field.

In case of medical diagnosis, data cannot be dealt with type-1 fuzzy set due to advanced directive of fuzziness involved with the information. So, type-2 fuzzy set is needed to be presented for attempting such information. If someone wants to operate all the methods of calculating the severity level of type-2 fuzzy set based medical diagnosis, then one can use the neural network amongst the basis and termini. Additional, if someone wants the controls of all the approaches without doing physically then one can generate the AI algorithm into neural network and make it artificial neural network. The basic objectives of the work are as follows:

a. Our objective is to use artificial intelligence (AI) especially in reasoning, in the development of medical diagnosis.
b. The type-2 fuzzy set will provide us an additional dimensionality, so that we are able to think in three-dimensional space, to deal the uncertainty.
c. To develop type-2 neuro fuzzy system by merging fuzzy inference system with neural network.
d. To optimize the parameters, involve in the output of the proposed algorithm. We apply the teaching learning-based optimization (TLBO).
e. The expected output of proposed system will give minimum error after applying the optimization tools for the known parameters.

2 Basic Concepts

2.1 Type-2 Fuzzy Set

A type-2 fuzzy set A on universal set X can be describe in such a way in which the membership value is again a fuzzy set, defined by:

$$A = \{((\tau, \nu), \mu_A(\tau, \nu)) : \tau \in X \text{ and } \nu \in I_\tau \subseteq [0, 1]\} \tag{1}$$

where, I_τ: Primary membership function and $\mu_A(\tau, \nu)$: S econdary membership function, also $0 \leq \mu_A(\tau, \nu) \leq 1$.

2.2 Standard Operations on Type-2 Fuzzy Set

Let $A_1 = \{((\tau, \nu_1), \mu_{A_1}(\tau, \nu_1)) : \tau \in X \text{ and } \nu_1 \in I_{\nu_1} \subseteq [0, 1]\}$ and

$$A_2 = \{((\tau, \nu_2), \mu_{A_2}(\tau, \nu_2)) : \tau \in X \text{ and } \nu_2 \in I_{\nu_2} \subseteq [0, 1]\}$$

are two type-2 fuzzy sets on universal set X. Then,

Union: $A_1 \cup A_2 \Leftrightarrow \{((\tau, \nu_1 \bigvee \nu_2), \mu_{A_1}(\tau, \nu_1) \star \mu_{A_2}(\tau, \nu_1))\}$
Intersection: $A_1 \cap A_2 \Leftrightarrow \{((\tau, \nu_1 \wedge \nu_2), \mu_{A_1}(\tau, \nu_1) \star \mu_{A_2}(\tau, \nu_1))\}$
Complement: $\overline{A_1} \Leftrightarrow ((\tau, 1 - \nu_1), \mu_{A_1}(\tau, \nu_1))$ and

$$\overline{A_2} \Leftrightarrow ((\tau, 1 - \varepsilon_2), \mu_{A_2}(\tau, \varepsilon_2))$$

where, $\star$ represents t-norm, $\bigvee$ indicates max t-co-norm, and $\wedge$ denotes min t-norm.

2.3 Artificial Neural Network (ANN)

ANN is commonly referred as **Neural networks** (NN). NNs are computing based structures encouraged by the biological NN that establish human brains. It is constituted with three layered structures "input layer", "hidden layer" and "output layer". The hidden layer may consist more than one layer. The various sections of the human brain are wired to proceed different types of information and the parts of brain are arranged in different layers. During the decision-making process, the information arrives in the brain, each layer does its specific job of processing the information, arising understandings, and transfer it into the next layer.

2.4 Uncertainty in Medical Diagnosis

In practice, the physicians routinely encounter diagnostic uncertainty. Despite its influence on health care management the measurement of diagnostic ambiguity is unwell understandable. To conduct a systematic view so that one can describe that how diagnostic ambiguity measured in medical training. The term "diagnostic ambiguity" lacks a clear meaning, and there is no complete framework for its dimension in medical field. The diagnostic ambiguity can be defined as a subjective perception of a failure to deliver an accurate description of the patient's health. The advancements in the measurement of diagnostic ambiguity can expand our understanding of diagnostic decision-making system to overcome the diagnostic errors.

3 Proposed Type-2 Fuzzy Set-Based Systems

In this section, we describe the architecture of type-2 neuro fuzzy system (Fig. 1) and Various components of the proposed system (Fig. 2).

3.1 Architecture of Proposed Type-2 Neuro Fuzzy System

See Fig. 1.

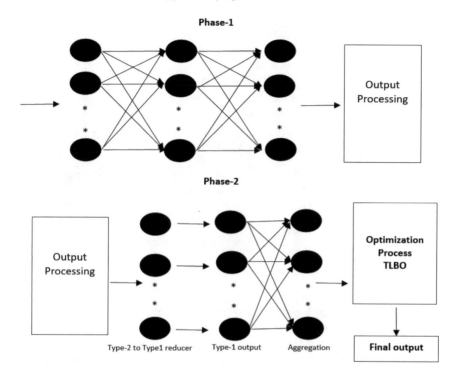

Fig. 1 Type-2 neuro fuzzy system

3.2 Architecture of Proposed System

See Fig. 2.

4 Algorithm of the Proposed System

Step 1: Consider the set of involved input factors;

$$I = \{I_1, I_2, \ldots\ldots\ldots I_n\} \tag{2}$$

Step 2: Applying fuzzification process to make type-2 fuzzy set for each input factors.

Step 3: Now, for the inference system process, we need to use some rules based on knowledge-based system. For this process, consider the conditional and unqualified intuitionistic fuzzy proposition for obtained type-2 fuzzy set input factors.

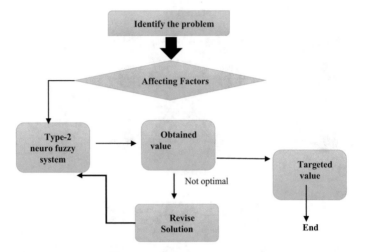

Fig. 2 Various components of the proposed system

Step 4: In this step the type-2 fuzzy set is converted into the form of type-1 fuzzy set by fuzzifying the area obtained in the 3rd direction of the type-2 fuzzy set.
Step 5: We aggregate the type-1 fuzzy set obtained by layer 4.
Step 6: In this study, "Sugeno's based inference system" is employed; we will do the defuzzification of the aggregated area obtained in layer 5.
Step 7: The final crisp output will be calculated by the following linear equation;

$$y = a_1 I_1 + a_2 I_2 + \ldots a_n I_n + c \tag{3}$$

where, a_i represents the weights.
Step 8: TLBO is used to optimize the unknown weights or parameters involve in the outcomes of neuro fuzzy system. TLBO works on the consequence of impact of an educator on the results of student. TLBO is a two-phased algorithm it consists:

(i) Teacher's phase
(ii) Learner's phase

In TLBO technique a group of students is measured as population and numerous topics offered to the students are measured as unlike project variable. The unlike variables are essentially the restrictions elaborate in the objective function. This procedure does not necessitate any 'algorithm-specific' control parameters.

Table 1 Input factors and their linguistic ranges

S. no	Symptoms	Linguistic ranges		
1	HRCT (CT)	<8 mild (M)	7–15 moderate (Md)	>14 risky
2	Blood level (BS)	<5 (M)	4–10 (Md)	>9 risky
3	X-ray level (XS)	<5 (M)	4–10 (Md)	>9 risky
4	U.G.C. level (UGC)	<5 (M)	4–10 (Md)	>9 risky

5 Data Collection

In this section, we have collected a secondary data collected from [21]. The data contained post-COVID 19 patients with HCRT thorax, blood test, chest X-ray, U.S.G. whole abdomen, kidney function test, liver function test and RTPCR report. Out of these factors, we have categories high-resolution CT (HRCT), blood severity, X-ray severity and U.G.C. severity have divided into three linguistic categories as shown in Table 1.

The output factors i.e., severity level is categories into five linguistic categories; low (0–0.2), M (0.2–0.4), Md (0.4–0.6), risky (0.6–0.8), and very risky (0.8–1).

6 Fuzzy Rules

The fuzzy rules for the type-2 neuro fuzzy system model are shown in Table 2, we have taken two input and one output-based neuro fuzzy system out of various rules, we have shown only 12 rules.

Table 2 Fuzzy rules foe the type-2 neuro fuzzy system

Rules	IF	AND	THEN
1	CT is M	BS is Md	Severity is Md
2	XS is risky	UGC is risky	Severity is very risky
3	UGC is M	BS is M	Severity is M
4	BS is risky	XS is risky	Severity is risky
5	XS is M	CT is Md	Severity is Md
6	CT is risky	BS is M	Severity is Md
7	BS is Md	XS is risky	Severity is risky
8	XS is M	UGC is Md	Severity is low
9	CT is M	CT is M	Severity is Md
10	CT is M	BS is Md	Severity is Md

7 Numerical Computation

Let us consider a patient with the input data.

CT	BS	UGC	XS
5	3	10	12

Based on the output obtained from Fig. 3, the final crisp output will be calculated by the following linear equation:

$$y = a_1 0.34 + a_2 0.967 + a_3 0.7 + a_4 0.967$$

where, a_i represents the weights. Let us consider the targeted values as 1. We now apply the TLBO technique to find the error of that system. TLBO is used to optimize the unknown weights or parameters involve in the outcomes of neuro fuzzy system.

From the Chart 1. We obtained the optimized values of the coefficients.

a_1	a_2	a_3	a_4	y
0.5828	0.5622	0.1813	0.1705	1.03

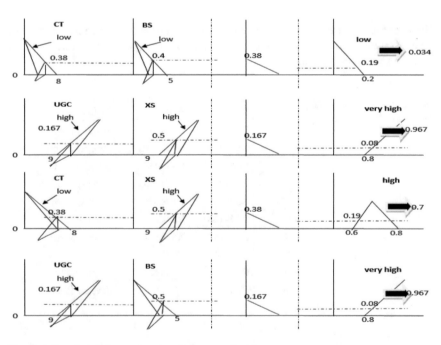

Fig. 3 Antecedent and consequent part of proposed system

The error of that system can be obtained by

$$y = \frac{|targated\ value - obtained\ valued|^2}{2} = \frac{|1.000 - 1.030|^2}{2} = 0.00045.$$

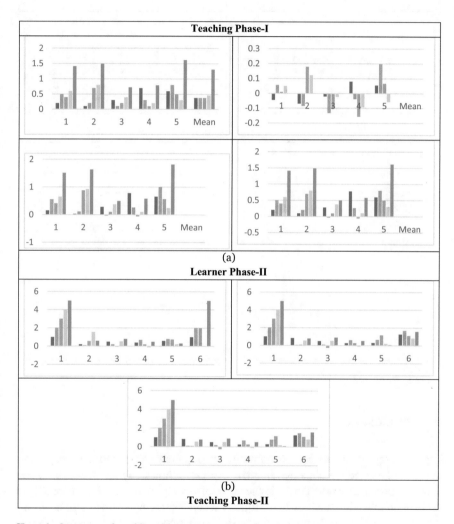

Chart 1 Outcomes of teaching phases (**a, c**) and learning phases (**c, d**)

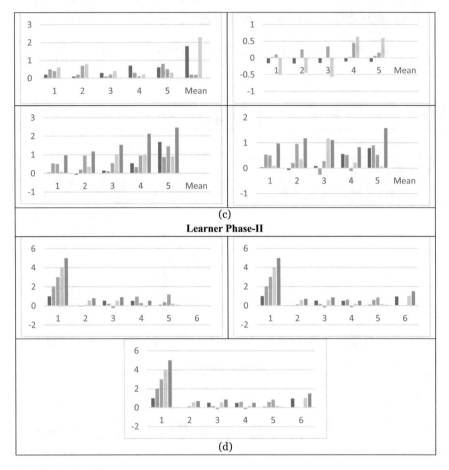

(c)
Learner Phase-II

(d)

Chart 1 (continued)

8 Conclusion

Type-2 fuzzy logic can be applicable in development of the neuro fuzzy system in the field of medical diagnosis in many aspects. While studying type-2 fuzzy sets uncertainty can be handled in 3-dimension figure by increasing the dimensionality of primary membership function in the form of secondary membership function. During this work, we have merged the type-2 fuzzy inference system with neural network. The weights used in the output of proposed neuro fuzzy system are handled by TLBO technique. The concept of type-2 fuzzy logic showed the impact of uncertain information with the help of neuro fuzzy system. Type-2 neuro fuzzy system have the ability of deal with the membership grade in higher dimensionality. Based on the proposed type-2 neuro fuzzy system, we can easily handle the doubtful situations that cannot be dealt with fuzzy logic or intuitionistic fuzzy logics. The applicability in the medical field of proposed neuro fuzzy system has given in this work by taking numerical example. From the numerical computation we got the error 0.00045 of the proposed system. For future perspective, we will try to develop a methodology to obtain an optimal process with the help of type-3 fuzzy logic that cover all the heterogeneous non-linear data of post COVID-19 patient.

Acknowledgements This work has been carried out under the University Grant Research Scheme Ref. Number Dev./1043 dated 29.06.2022. The first author is thankful to UGC for financial assistance.

References

1. Zadeh, L.A.: Fuzzy sets. Inf. Control **8**(3), 338–353 (1965)
2. Zadeh, L.A.: The concept of a linguistic variable and its application to approximate reasoning—I. Inf. Sci. **8**(3), 199–249 (1975)
3. Xie, B.K., Lee, S.J.: An extended type-reduction method for general type-2 fuzzy sets. IEEE Trans. Fuzzy Syst. **25**(3), 715–724 (2017)
4. Castillo, O., Melin, P.: Adaptive noise cancellation using type-2 fuzzy logic and neural networks. In: 2004 IEEE International Conference on Fuzzy Systems (IEEE Cat. No. 04CH37542) **2**, 1093–1098. IEEE (2004)
5. Castillo, O., Huesca, G., Valdez, F.: Evolutionary computing for optimizing type-2 fuzzy systems in intelligent control of non-linear dynamic plants. In: NAFIPS 2005–2005 Annual Meeting of the North American Fuzzy Information Processing Society, pp. 247–251. IEEE (2005)
6. Coupland, S., John, R.: A new and efficient method for the type-2 meet operation. In: 2004 IEEE International Conference on Fuzzy Systems (IEEE Cat. No. 04CH37542) **2**, 959–964. IEEE (2004)
7. Hagras, H.A.: A hierarchical type-2 fuzzy logic control architecture for autonomous mobile robots. IEEE Trans. Fuzzy Syst. **12**(4), 524–539 (2004)
8. Hagras, H.: A type-2 fuzzy logic controller for autonomous mobile robots. In: 2004 IEEE International Conference on Fuzzy Systems (IEEE Cat. No. 04CH37542) **2**, 965–970. IEEE (2004)
9. Hwang, C., Rhee, F.C.H.: An interval type-2 fuzzy C spherical shells algorithm. In: 2004 IEEE International Conference on Fuzzy Systems (IEEE Cat. No. 04CH37542) **2**, 1117–1122. IEEE (2004)
10. Figueroa, J., Posada, J., Soriano, J., Melgarejo, M., Rojas, S.: A type-2 fuzzy controller for tracking mobile objects in the context of robotic soccer games. In: The 14th IEEE International Conference on Fuzzy Systems. FUZZ'05, pp. 359–364. IEEE (2005)
11. Garibaldi, J.M., Musikasuwan, S., Ozen, T.: The association between non-stationary and interval type-2 fuzzy sets: a case study. In: The 14th IEEE International Conference on Fuzzy Systems. FUZZ'05, pp. 224–229. IEEE (2005)
12. Lin, P.Z., Hsu, C.F., Lee, T.T.: Type-2 fuzzy logic controller design for buck DC–DC converters. In: The 14th IEEE International Conference on Fuzzy Systems. FUZZ'05. pp. 365–370. IEEE (2005)
13. Lynch, C., Hagras, H., Callaghan, V.: Embedded type-2 FLC for real-time speed control of marine and traction diesel engines. In: The 14th IEEE International Conference on Fuzzy Systems. FUZZ'05, pp. 347–352. IEEE (2005)
14. Mittal, K., Jain, A., Vaisla, K.S., Castillo, O., Kacprzyk, J.: A comprehensive review on type 2 fuzzy logic applications: Past, present and future. Eng. Appl. Artif. Intell. **95**, 103916 (2020)

15. Rhee, F.C.H., Hwang, C.: A type-2 fuzzy C-means clustering algorithm. In: Proceedings Joint 9th IFSA World Congress and 20th NAFIPS International Conference (Cat. No. 01TH8569) **4**, 1926–1929. IEEE (2001)
16. Rhee, F.H., Hwang, C.: An interval type-2 fuzzy perceptron. In: 2002 IEEE World Congress on Computational Intelligence. 2002 IEEE International Conference on Fuzzy Systems. FUZZ-IEEE'02. Proceedings (Cat. No. 02CH37291) **2**, 1331–1335. IEEE (2002)
17. Sepúlveda, R., Castillo, O., Melin, P., Rodríguez-Díaz, A., Montiel, O.: Integrated development platform for intelligent control based on type-2 fuzzy logic. In: NAFIPS 2005–2005 Annual Meeting of the North American Fuzzy Information Processing Society, pp. 607–610. IEEE (2005)
18. Ontiveros-Robles, E., Melin, P., Castillo, O.: Comparative analysis of noise robustness of type 2 fuzzy logic controllers. Kybernetika **54**(1), 175–201 (2018)
19. Ruiz-García, G., Hagras, H., Pomares, H., Ruiz, I.R.: Toward a fuzzy logic system based on general forms of interval type-2 fuzzy sets. IEEE Trans. Fuzzy Syst. **27**(12), 2381–2395 (2019)
20. Sennan, S., Ramasubbareddy, S., Balasubramaniyam, S., Nayyar, A., Abouhawwash, M., Hikal, N.A.: T2FL-PSO: type-2 fuzzy logic-based particle swarm optimization algorithm used to maximize the lifetime of Internet of Things. IEEE Access **9**, 63966–63979 (2021)
21. Sharma, M.K., Dhiman, N., Mishra, V.N., Mishra, L.N., Dhaka, A., Koundal, D.: Post-symptomatic detection of COVID-2019 grade based mediative fuzzy projection. Comput. Electr. Eng. **101**, 108028 (2022)

A Survey on Type-2 Fuzzy Logic Systems in Healthcare

Sindhu Rajendran, B. Sahana, Shaik Farheen, B. N. Meghana, and Sona Theresa Babu

Abstract Fuzzy logic plays a key role in various areas, particularly medicine. One of the major problems faced by developed and developing countries is medical treatment. There has been a great deal of concern in finding an alternative to traditional methods of diagnosis due to imprecision and inaccuracy. The fuzzy logic has made identification and diagnosis process rapid, thus minimizing the intervention time and treatment methods with greater accuracy. In fuzzy logic, type 2 logic systems and type 2 Fuzzy-based methods are preferred since they offer the major advantage of demonstrating a variety of outputs and provide a powerful framework for reasoning. This chapter includes the study of type 2 fuzzy logic systems used for treating particular diseases and their role in the development of the medical diagnostic system. The first aim broadly discusses how fuzzy logic has evolved over the years and its need in different sectors. The second aim is to analyze the applications of type 2 fuzzy logic in various branches and discusses recent developments that have occurred mainly in the medical sector. The third aim is to include type 2 fuzzy systems that are used to treat particular diseases paving the way to find further scope for further developments.

Keywords Type-2 fuzzy logic · Type-2 fuzzy-based methods · Disease diagnosis

S. Rajendran (✉) · B. Sahana · S. Farheen · B. N. Meghana · S. T. Babu
R. V. College of Engineering, Bengaluru 560059, India
e-mail: sindhur@rvce.edu.in

B. Sahana
e-mail: sahanab@rvce.edu.in

S. Farheen
e-mail: shaikfarheen.ec20@rvce.edu.in

B. N. Meghana
e-mail: bnmeghana.ec20@rvce.edu.in

S. T. Babu
e-mail: sonathereseb.ec20@rvce.edu.in

© The Author(s), under exclusive license to Springer Nature Switzerland AG 2023
O. Castillo and A. Kumar (eds.), *Recent Trends on Type-2 Fuzzy Logic Systems: Theory, Methodology and Applications*, Studies in Fuzziness and Soft Computing 425,
https://doi.org/10.1007/978-3-031-26332-3_11

165

1   Introduction

The word Fuzzy means inexact or imprecise. Fuzzy logic was first conceptualized by *Lofti A Zadeh in 1973*. Fuzzy logic is a branch of artificial intelligence and a concept with numbers where the true values of variables can be any real number between 0 and 1 [1]. In contrast to other traditional logical systems, Fuzzy logic provides a mean to amend reasoning which is approximate rather than accurate, especially human reasoning. It uses a logic and decision mechanism which has no boundaries rather than binary logic (0 or 1) or human logic. From recognizing handwriting to detecting cancer, fuzzy logic has simplified and improved effective methods. Any system working on logic can be evaluated by fuzzy logic. Based on web applications and interfaces (even in the absence of an expert) users can expect results from the web applications which work on an inbuilt knowledge base. It can be concluded that fuzzy logic can represent human intelligence, unlike Boolean or crisp logic [2].

Fuzzy logic systems manipulate membership functions (these are functions that measure the degree of truth in fuzzy sets) which simulate variables with a rule-based inference engine. In a fuzzy set, its elements are all mapped in between [0, 1] and are characterized by membership functions. Fuzzy logic has two categories-T-1 and T-2. Type-2 fuzzy logic (T-2 FL) was conceptualized in 1973. Type 1 fuzzy set (T-1 FS) is employed to find the extent of achieving the features of the object, whereas Type-2 Fuzzy set (T-2 FS) cannot determine the extent of achieving the characteristics of the object [3].

Type-2 fuzzy logic system (T-2 FLS) has uncertainty in its membership function also. T-2 FL is an inference generalization of the conventional type 1 fuzzy logic (T-1 FLS). Membership functions in T-2 FLS are T-2 FS's. Knowledge utilized in assembling rules for fuzzy logic systems is often uncertain. This uncertainty leads to uncertainty in consequent membership functions, which cannot be operated by T-1 FS. T-2 FS's are beneficial in deciding the exact membership function of fuzzy logic sets, the membership grades of T-2 FS's are T-1 FS's [4].

Abbreviations

Denoted as	Full form
T-1 FLS	Type 1 Fuzzy Logic System
T-2 FLS	Type 2 Fuzzy Logic System
T-1 FS	Type 1 Fuzzy Set
T-2 FS	Type 2 Fuzzy Set

1.1 Fuzzy Logic Versus Boolean Logic System

Every output has a degree. A precise hypothesis is considered the limiting case of an estimated hypothesis. Fuzzy logic is a concept that depends on the degrees of truth and is based on possibility theory whereas traditional logic systems like Boolean logic are dependent on true or false and probability theory. Here is an example comparing fuzzy logic and Boolean logic.

Boolean/binary logic: The logic that generates a single value representing either 0 or 1; on or off; true or false; yes or no as output is called Boolean/binary logic.

In Boolean logic, truth is a result of reasoning obtained from inexact or partial knowledge.

Fuzzy logic: It is a concept of partial truth ranging between completely true and completely false.

In fuzzy logic truth is a result of recognizing, interpreting data and information which is imprecise and non-numerical.

According to the Aristotelian logic, either the person is perfectly all right (healthy) or the person is not at all healthy (probably dead).

Let 'p' be the degree that reflects his/her health condition,

$p = 1$ (it denotes that person is healthy).

$p = 0$ (it denotes that he/she is not healthy at all and may be suffering from terminal cancer).

The same matter when discussed using fuzzy logic, there comes another factor that is to be taken into consideration 'i'.

i.e., $p + i = 1$.

Here, 'i' denotes the degree of illness.

In extreme conditions, $p = 0$ and $i = 1$ (a person is not at all healthy).

For a small injury, $p = 0.9999$, and $i = 0.0001$.

Painful ulcer, $p = 0.45$ and $i = 0.55$.

Terminal cancer, $p = 0.05$ and $i = 0.9$.

This shows that Boolean Logic is a subset of Fuzzy Logic.

1.2 T-1 FLS Versus T-2 FLS

A T-2 FS is specified by a fuzzy membership function, i.e., the membership grade of each element that belongs to the T-2 FS is a fuzzy set in the interval 0–1, whereas in T-1 FS it is a crisp number in [0, 1]. In many circumstances that we come across in real life, T-1 FS is examined as the first-order estimation whereas T-2 FS is examined as the second-order estimation [5]. Uncertainty is only restricted to lingual parameters in T-1 FL while in T-2 FL it is also present in membership functions (Fig. 1).

Comparative study onT-1 FLS and T-2 FLS [7].

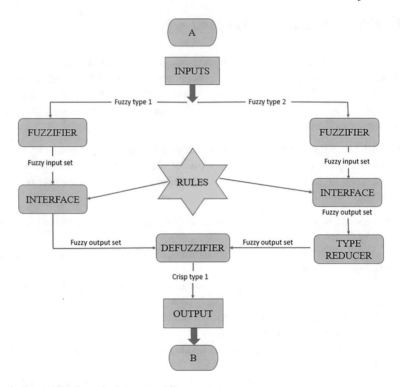

Fig. 1 T-1 and T-2 fuzzy logic systems [6]

(1) Classification accuracy and noise resistance

The noise resistance and accuracy of T-2 FLS are more advanced than that of T-1 FLS.

(B) Performance

Users are more satisfied with the outputs of the T-2 FLS than T-1 FLS due to its highly accurate results in disease diagnosis.

(C) Handling uncertainties

T-2 FLS is more efficient in operating uncertainties and in the classification of imprecision data compared to T-1 FLS.

(D) Prediction

T-2 FLS overrules T-1 FLS in prediction as there are more rules in T-2 when compared with T-1.

1.3 Need for T-2 FLS

Digital transformation in all sectors all over the world is rapidly advancing with time. There is a need for a logic system that deals with human uncertainties and predicts risks sooner (be it in the medical field or industrial sector, or business). One such intelligence that evaluates every logic-related system is Fuzzy logic. It works on an extent of probabilities of inputs to produce a particular output.

Fuzzy logic acknowledges human emotions through conversion analysis. It assigns truth values like true, partially true, likely, untrue, and very true to linguistic variables leading to the desired fuzzy output. Since uncertainty is common in fields such as medicine, fuzzy logic assumes such uncertainty using fuzzy set theory, and other fuzzy methods derived from fuzzy set theory can be considered legitimate objects to deal with such uncertainty. As fuzzy logic deals with uncertainties and possibilities to obtain accurate conclusions, it helps in diagnosis, monitoring the patient, and measuring the extent of the danger of diseases in medical applications. To increase the diagnosis of a disease, T-2 FLS is much more capable than T-1 FLS when uncertainty modeling is used. Since T-2 FS has more parameters and the membership degree of each element in sets will be a fuzzy set itself, uncertainties are overseen (Table 1).

The applications of T-2 FL methods in the medical field are mentioned in Table 2.

2 Methodology of Diagnosing Disease Using T-2 FL

The Fuzzy logic is preferred when there are a lot of segments of information and when they are disposed to scalar quantities. Figure 2 shows the approach to finding the probability of disease using scalar quantities in the medical field [6].

Step 1: Selecting the Proper Fuzzy Method
Every method doesn't suit a particular disease.
Ex-FES for Chronic kidney disease and spinal cord disorders are diagnosed by a rule-based type-2 fuzzy system.
Step 2: Define and Model Disease
Analyzing the disease is the main step to knowing about the inputs to be given so that we end up with the expected output.
Step 3: Assigning Input Variables
From step 2 we get to know the inputs to be used. In this step, symptoms of the particular disease are given as inputs.
Step 4: Fuzzifying Input Value.
In this step, the input values are charted from 0 to 1 using input membership functions.
Step 5: Executing Applicable Methods to Compute Fuzzy Output Functions

Table 1 Applications of type 2 fuzzy logic in various fields [8]

Sl no.	Fields	Applications
1	Education field	E-learning evaluation, academic progress of a student, learning outcomes of a student
2	Wireless communications	Decoding, estimating channel, channel equalization
3	Architecture	Computes worth of housing locations on site of plan, also gives us an outcome of most profitable to the least
4	Business, management field	Computing, documentation reduces and save budget
5	Defense field	The automatic target recognition was used by the North Atlantic treaty of organization
6	Psychology	Criminal investigation and analysis of human behaviour
7	Security system	The proper security system in trading
8	Finance field	Used in fund management and to make predictions in Stock market
9	Medical field	Monitoring patients, diagnosis of diseases, decision-making in radiotherapy, etc.
10	Manufacturing industry	Enhancement of milk and its products (cheese, yogurt, butter) manufacturing
11	Aerospace	Altitude control in aircraft and satellites
12	Automotive and transportation	Speed and acceleration control, braking of automotive (Like train, bus, car), traffic control
13	Industrial field	Applied in wastewater treatment, and activated sludge treatment

Based on the interactions between rule bases and inference engine from knowledge base output will be computed. The output of the T-2 FLS will be a T-2 FS which will be type-reduced to type-1 by type-reduction.

Step 6: Defuzzifying the Output

Defuzzification is a process of finding a key number as an output i.e., it is used to fetch the fuzzy results into a single number/crisp number. Type-reduced set will be defuzzified into a crisp number (type-0).

Step 7: Producing Output

In this step, we get the disease as the output.

2.1 Type-2 Fuzzy Systems

The basic rules of inference continue to apply in any type-n fuzzy logic. The degree of fuzziness increases as n in type-n increases. It changes the characteristics and operations of membership functions but basic principles remain the same. The only difference between concepts of T-1 and T-2 FL is in relation to the characteristics of membership functions. The output sets of T-1 fuzzy systems are T-1 FS. Then

Table 2 Applications of T-2 FLS in the medical field [9]

Sl no.	Disease	Diagnosis
1	Diabetes	Utilizing statistical data and the framework of the T-2 fuzzy neural system blood glucose levels of the patient can be controlled. The fuzzy inference was used for the diagnostic program of diabetic patients by dynamical dose analysis answers to glucose tolerance tests
2	Heart disease	T-2 fuzzy expert system is used to diagnose a possible heart disease for a patient. One of the applications is the use of incomprehensible control of the artificial heart as a whole. Maintaining sufficient organ filling by controlling the pumping rate is the main function of the artificial heart control system
3	Lung cancer	Risk factors and symptoms of lung cancer are gathered to generate rules of the fuzzy logic system and are stored in the rule base. T-2 fuzzy inference engine fires irrelevant rules and output is delivered as the probability of disease
4	Leukemia	T-2 FL is implemented in designing Fuzzy expert system for treatment of lymphocytic leukaemia [10]
5	Bechterew's disease	T-2 fuzzy rule-based expert system is operated in diagnosing Bechterew's disease (a type of arthritis). T-2 FL manages the ambiguity related to lingual parameters in the rules and indefinite numeric inputs in the whole diagnosis process
6	Asthma	T-2 fuzzy logic expert system is employed to check the severity level in the diagnosis of asthma and chronic stage pulmonary diseases
7	Depression	T-2 fuzzy expert system is employed in the diagnosis of depression as a demanding aspect in the treatment of depression is to reduce time loss and increase accuracy
8	Malaria	T-2 FLS builds its rules and sets the fuzzy logic provisions from the data available and the same is used in the diagnosis of malaria
9	Spinal cord disorders	Hybrid rule-based T-2 FLS is utilized to supervise the high variability of diagnosing chronic disorders and severity
10	Paediatric	The fuzzy logic is accustomed to analyzing the pattern and interpretation of changes in the cardiotocograph (pattern of foetal heart rate and uterine contractions) during childbirth that may lead to inessential or lethal medical intervention damage
11	Oncology	The utility of the fuzzy logic concept in oncology has been associated with the function of segregation (discrimination of common tissue from cancer) and treatment advice function

(continued)

Table 2 (continued)

Sl no.	Disease	Diagnosis
12	Gerontology	The fuzzy logic concept is used for clustering in gerontology
13	Neurology	Fuzzy logic and neural networks find their prime roles in neurology. Such as signal and image processing, clustering analysis, etc.
14	Radiation	Diagnostic imaging and radiation therapy planning have become very popular topics in this field that uses fuzzy logic
15	Odontology	A Fuzzy logic method called the Fuzzy expert system helps to keep a record of dental radiographs and automatic polishing of Cobalt chromium
16	Malignant tumor of the breast	Fuzzy logic methods have a pre-diagnosed malignant tumor of the breast (breast cancer) by extracting smears of breast mass and sorting data with a sensitivity of 98.6% which is integral in the early treatment of such a deadly disease

Fig. 2 Flowchart for diagnosing disease using T-2 FL [6]

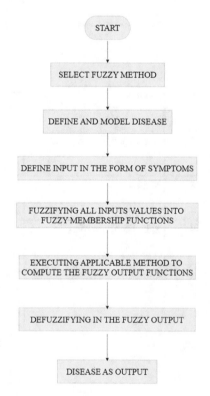

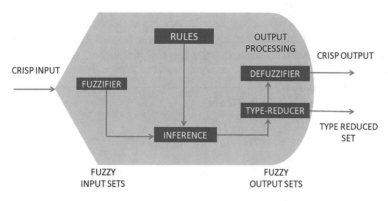

Fig. 3 Type2 fuzzy system architecture [11]

Defuzzification is done to obtain output that must be a crisp illustrative of all the integrated output sets. The output sets of T-2 fuzzy systems are-2 FS. A crisp number is obtained as the product of a fuzzy system on the Defuzzification of T-1 whereas a T-1 FS is obtained as an output of a fuzzy system on the Defuzzification of type-2. This performance of type-2 Defuzzification is known as "type-reduction" and the fuzzy set acquired at the output which is of type-1 is called the "type-reduced set". This type-reduced fuzzy set i.e., T-1 FS is again defuzzied to obtain a crisp number as the product of the T-2 FLS [11] (Fig. 3).

2.2 Operations on T-2 FL

T-2 FS, $\tilde{A}_i$ (i = 1, 2...., r) have secondary membership functions as T-1 FS. Binary operations of minimum and maximum, and the unary operation of negation, are to be extended from crisp numbers to type-1 fuzzy sets to enumerate the union, intersection, and complement of T-2 FS, as for each x, $\mu_{\tilde{A}_i}(x, u)$ is a function (where $\mu_{\tilde{A}_i}(x)$ is a crisp number in the T-1 FS) [11]. Let the T-2 FS be $\tilde{A}_1$ and $\tilde{A}_2$, i.e.

$$\tilde{A}_1 = \int_x \mu_{\tilde{A}1}(x)/x \qquad (1)$$

$$\tilde{A}_2 = \int_x \mu_{\tilde{A}2}(x)/x \qquad (2)$$

The union of T-2 FS: $\tilde{A}$
The $\cup$ of $\tilde{A}_1$ and $\tilde{A}_2$ gives A T-2 FS.

$$\tilde{A}_1 \cup \tilde{A}_2 = \int_x \mu_{\tilde{A}_1 \cup \tilde{A}_2}(x)/x \qquad (3)$$

Typically, a union of two secondary membership activities is formed between all potential pairs of primary membership. If one or more pairs of all potential pairs give the same output (point), it means that the one with the highest membership rate is taken [11].

The Intersection of T-2 FS.

The $\cap$ of $\tilde{A}_1$ and $\tilde{A}_2$ gives A T-2 FS.

$$\tilde{A}_1 \cap \tilde{A}_2 = \int_x \mu_{\tilde{A}_1 \cap \tilde{A}_2}(x)/x \tag{4}$$

The discrepancy between the two functions of secondary membership is made between all potential pairs of primary membership. If one or more pairs of all potential pairs give the same output (point), it means that the one with the highest membership rate is taken [11].

Complement of T-2 FS.

The complement of set $\tilde{A}$ gives a T-2 FS.

$$\tilde{A}' = \int_x \mu_{\tilde{A}'_1}(x)/x \tag{5}$$

In Eq. (5) $\mu_{\tilde{A}1}$ is a secondary membership function, that is, for every value of x, $\mu_{\tilde{A}1}$ is a function (contrary to T-1 case where, at every value of x, $\mu_{\tilde{A}1}$ is a point value) [12].

3 Type Reduction

In T-1 FLS, the output relative to rules is a T-1 FS in space. The defuzzifier combines all the type-1 output sets in certain ways to obtain a crisp output that represents the combined output sets. For example, the centroid defuzzifier gives the centroid of the fusion of all the output T-1 FS as the crisp output. InT-2 FLS, the result relative to rules is a T-2 FS in space. The type-reducer in fuzzy system integrates the output sets in different ways just as the T-1 defuzzifier and centroid computation results in a T-1 FS which is called a "type-reduced set" [6].

3.1 Centroid Type Reduction

In the case of T-1 FLS, the centroid defuzzifier integrates all the output sets in T-1, the centroid of this set will be crisp output. Let the combined output fuzzy set of all output sets be B, the centroid defuzzifier is mentioned in Eq. (6).

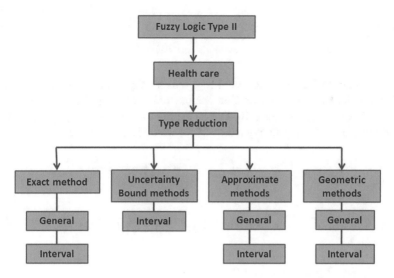

Fig. 4 Framework of type-2 reduction in healthcare [6]

$$y_c(x) = \frac{\sum_{i=1}^{N} y_i \mu_B(y_i)}{\sum_{i=1}^{N} \mu_B(y_i)} \tag{6}$$

where, the output set B is summed to N points.

The centroid type-reducer comes into play in combining output sets of T-2. It integrates all the product sets of T-2 by computing their union. The expression for the centroid type-reduced set is an elongation of the centroid defuzzifier of T-1 FS (Eq. 6). The final expression of the type-reduced set is given in Eq. (7) (Fig. 4).

$$\tilde{Y}_c(x) = \int \Theta 1 \ldots \ldots \int \Theta N \; \tau_{l=1}^{N} \; \mu_{Di}(\Theta_i) \; \frac{\sum_{i=1}^{N} y_i \Theta_i}{\sum_{i=1}^{N} \theta_i} \tag{7}$$

where $\{\Theta_1, \ldots, \Theta_N\}$ are such that $\sum_{i=1}^{N} y_i \Theta_i \sum_{j=1}^{N} y_j \Theta_j = y$.

3.2 Height Type Reduction

In the case of T-1 FLS, the height defuzzifier restores each output of a T-1 FS with a singleton set at the point where that output set has maximum membership and so the centroid of all the output sets of T-1 comprising singletons is computed. The product of a height defuzzifier is mentioned in Eq. (8).

$$Y_h(x) = \frac{\sum_{l=1}^{M} y^{-1} \mu_B(y^{-1})}{\sum_{l=1}^{M} \mu_B(y^{-1})} \tag{8}$$

When it comes to T-2 FLS, the height type-reducer supplants each output set of T-2 with a T-2 singleton set whose region features only one point, i.e., the membership of which may be a type-1 set in [0, 1]. The expression obtained for the type-reduced set within the T-2 FLS is an elongation of (8) which is given as (9).

$$\tilde{Y}_h(x) = \int_{\Theta 1} \cdots \int_{\Theta M} \tau_{l=1}^{M} \mu_{D1}(\Theta_1) \frac{\sum_{l=1}^{M} y^{-1} \Theta_1}{\sum_{l=1}^{M} \theta_1} \tag{9}$$

where $\{\Theta_1, \ldots, \Theta_M\}$ are such that $\sum_{l=1}^{M} y^{-1} \Theta_1 \sum_{l=1}^{M} /y^{-1} \Theta_1 = y$.

3.3 General and Interval T-2 FLS

A complex T-2 interval fuzzy set is a fuzzy set where the membership grade of all domain points is a crisp set at any specified intervals set in [0, 1]. Interval T-2 FS require much simpler mathematics i.e., primarily interval arithmetic compared to general T-2 FS. We use general T-2 or interval T-2 reckoning on the amount of uncertainty and complexity [13].

Moderate uncertainty and non-critical problems: for instance, for classifying flowers, IT2 fuzzy logic is usually recommended (Fig. 5).

Moderate uncertainty and critical problem: as an example, for medical applications, GT2 fuzzy logic is usually recommended.

High uncertainty in data: GT2 fuzzy logic is usually recommended.

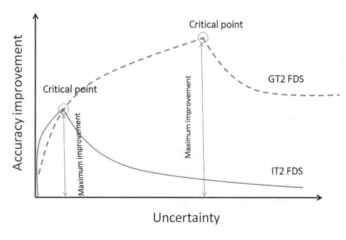

Fig. 5 GT2 and IT2 FDS Behavior accuracy improvement versus uncertainty [13]

4 T-2 FL Techniques in the Development of the Medical Field

T-2 FL is highly suitable in the medical domain. Therefore, the results of the research study vali-dated the effectiveness of the use of fuzzy meth-ods in the diagnosis process, presenting new in-formation to researchers about the type of disease that was the main focus. This will help determine the diagnostic features of the neglected medical fields. In the medical field, these tasks have to be accomplished for getting outputs on any input of medical data in any fuzzy logic system [14]. Outputs of different fuzzy tasks i.e., Clustering, Ranking analysis, Pattern Recognition, Classification, Fea-ture extraction, and Prediction are shown in Fig. 6.

Clustering: Clustering is an output of a set of data (containing all sorts of cells and genes), where the data is grouped into different sets called clusters depending on the properties of the cells and genes. All the cells in one cluster will be diagnosed as one disease while different clusters are not related to each other.

Prediction: Prediction deals with analyzing data at present using techniques like artificial intelligence and statistics to predict future complications. For example, the type-2 fuzzy rule-based system has improved prediction accuracy to predict breast cancer.

Pattern recognition: Recognizing regularities and patterns in a sample of data, so that irrelevant patterns can be removed and the decision-making process is simplified.

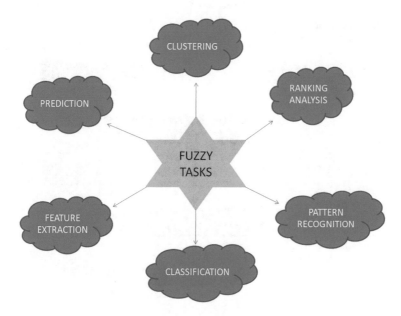

Fig. 6 Outputs of different Fuzzy tasks [4]

Classification: Classification accords with membership function which includes an individual of a particular class (elements that have the same property).

Ranking analysis: Ranking analysis is about ranking alternatives from best to worst. In the medical field, alternatives are criteria we include in the decision-making process. For example, the health performance of the patient and the risk factor of the disease.

4.1 T-2 Fuzzy Rule-Based System and Its Role in Diagnosis of Spinal Cord Disorders

It is most commonly used fuzzy inference system for a variety of problems. The framework of the T-2 fuzzy rule-based system is identical to that of the T-1 fuzzy logic system.

Figure 7 explains the architecture of the T-2 fuzzy rule-based system.

It has four elements [15]:

(1) User interface.
(2) Knowledge base.
(3) Inference engine.
(4) Working memory.

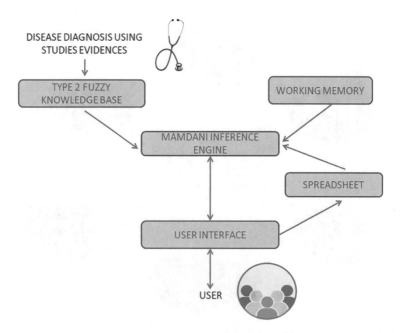

Fig. 7 Architecture of fuzzy T-2 rule-based system [15]

A rule-based system can either be a forward-chaining approach or a backward-chaining approach. Initially, information is gathered from the data that we get from study shreds of evidence and that information is stored in working memory. The rules and parameters that are required by the information gathered are analyzed with the material present in working memory. These rules are then analysed using an inference engine so that we end up with fuzzy results. Using the user interface, the fuzzy results could be depicted in different forms, most preferably the graphical user interface [15].

1. **User interface**: This enables humans to interact with the expert system.
2. **Fuzzifier**: A set of crisp input given by the user is converted into fuzzy sets based on inference rules from the knowledge base.
3. **Knowledge base**: To allow appropriate reasoning, it is necessary for any fuzzy logic system to store the rules and relations of linguistic variables derived from fuzzy set theory, so the knowledge base of FES represents this task.
4. **Inference engine**: The Inference engine is accountable for generating output by sending the inference inputs from the fuzzifier to fuzzy input. It estimates linguistic values to map them accordingly to fuzzy sets.
5. **Defuzzifier**: From a fuzzy output, realized using a decision-making algorithm, it is converted into a single crisp value, understandable by a human.

Diagnosing spinal cord disorders—In the modern day, a good chunk of our population faces problems related to our spinal cord. T2 rule-based system is used to handle the high uncertainty of disease diagnosis and gives the most appropriate result. It uses the method of forward and backward chaining for diagnosis, considering the most common spinal cord disorders—spinal stenosis and disc herniation.

Figure 8 gives the framework of forward chaining phase. If the primary problem of the patient is pain in the leg and the lower back or in the arm and neck, the system activates the knowledge base of the module of disc herniation while if the primary problem is a pain in both legs and both arms, the system activates the knowledge base of the module of spinal stenosis. The knowledge base of each module consists of the rules regarding the severity of pain in a specific location, the initial time of pain, and the dependence of pain on certain conditions.

Figure 9 depicts the mechanism of the backward chaining phase. The system tries to investigate some clinical symptoms to prove the primal diagnosis. The maximum value among the two primal diagnoses specifies the direction to select the next module. If the value of herniated disc disorder is maximum, the system activates the module of the nerve root to assure itself of the diagnosis and recognizes the compressed nerve root and the exact location of the abnormal disc.

4.2 T-2 Fuzzy Neural Systems

T-2 FLS in medical fields often want to set healthy lifestyles in patients by diagnosing disease thoroughly and springing up with appropriate diet and medicine or treatment

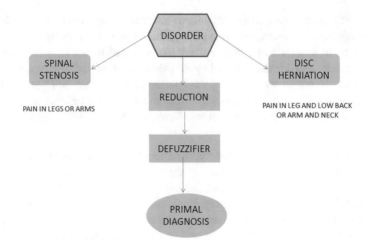

Fig. 8 Forward chaining phase [16]

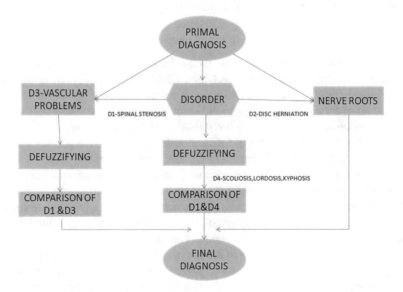

Fig. 9 Backward chaining phase [16]

to be followed. Diagnosis of diabetes may well be done using T-2 fuzzy neural systems. T-2 fuzzy neural network-based identification system uses two datasets in diagnosing diabetes. T-2 fuzzy neural network is an integrated combination of the T-2 fuzzy inference scheme and neural network structure. The accuracy of neural network T-2 FLS can be increased by considering different fuzzy rules [17] (Fig. 10).

The first layer includes the input layer which distributes the signal. Layer 2 includes T-2 membership functions utilized in mentioning undetermined terms. The

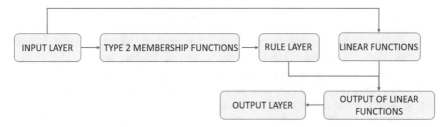

Fig. 10 Flowchart of neural network of T-2FLS used treating diabetes [17]

third layer includes the rule layer which consists of the fuzzy rule used for diagnosis. The Fifth and sixth layers execute type reduction and defuzzification functions. The inference engine is employed to seek out the crisp output that helps in diagnosing diabetes [17].

Fuzzy Neural System Application to Diagnose Erythemato-Squamous Diseases

In dermatology, the differential diagnosis of erythemato-squamous diseases is challenging due to the main reason stating different classes of the disease share the clinical features with very few differences. Therapeutically, patients are first assessed with regard to 12 features, which include a degree of scaling and erythema; itching; formation of papule; if lesion borders are present or absent; genealogicalchart; and involvement of the oral mucosa, knees, elbows, and scalp, which are important symptoms in the differential diagnosis of Erythemato-squamous diseases. The fuzzy neural system is advantageous in the aid of determining skin type, differential diagnosis, and treatment decision-making [18].

Method of diagnosing Erythemato-squamous diseases.

Analysis of dataset

In dermatology, Erythemato-squamous disease diagnosis is a difficult mission because all the six classes (pityriasisrubra, seborrheic dermatitis, psoriasis, lichen planus, chronic dermatitis, and prosea) share almost common clinical features with slight deviations. A patient with a specific disease may have the symptoms of another disease at a very initial stage and characteristic features of the following stages, which is another complication faced by dermatologists when accomplishing the differential diagnosis of these diseases. On analysing the samples under a microscope, the histopathological features were deduced. The genealogical chart feature in the dataset to be assembled may take the value '1' if any of these diseases have been diagnosed in the family and the value '0' if none in the family is diagnosed. Age features can be indicated by the patient's age. All other features (clinical and histopathological) can be attributed to values ranging from 0 to 3 (0 indicating absence of features; 1 and 2 indicating relative intermediate values; 3 indicating highest amount).

Proposed Algorithm [Fuzzy Neural Network (FNN)]

Fuzzy logic architectures are utilized in solving several complexities, which include categorization, identification, and detection. Among known convenient and easy methods for mapping an input to the corresponding output, the easiest approach is fuzzy logic. This mapping of input to output is attained using the if–then rule. Generally, linguistic terms are used to detail the values of variables in the fuzzy system. A membership function in a fuzzy system characterizes each fuzzy value. Implementing fuzzy logic reduces the complexity of data and handling of uncertainty and imprecision. The FNN, a combination of fuzzy logic and Neural networks allows us to map and design nonlinear systems. The dataset analysis obtained represents the FNN classifier's input signals. These input signals are again classified into six specified classes. The input signals of the system are evenly distributed in the first layer of FNN. The second layer consists of membership functions describing linguistic terms.

The nonlinearity and linearity approximated via fuzzy and neural network systems has this structure:

$$y_j = b_j + \sum_{i=1}^{m} a_{ij} x_i \tag{10}$$

x_i represents the input signals of system. y_i represents the output signals of the system. $i = 1,\ldots,m$ denotes the number of input signals while $j = 1,\ldots,r$ represents the number of rules. b_j and a_{ij} as the coefficient of the fuzzy set input. In the third layer, the rule layer is mounted where number of nodes is made equal to number of rules. The output signals of third layer are to be multiplied by the output signals of the fourth layer inducing the fifth layer. The Fuzzy Neural Network output signals enumerated in the sixth layer as shown below:

$$u_k = \frac{\sum_{j=1}^{r} W_{jk} y_j}{\sum_{j=1}^{r} \mu_j(x)} \tag{11}$$

u_k denotes the output signals of network system ($k = 1,\ldots,n$).

4.3   T-2 Fuzzy Expert System

Expert systems require comprehensive data about a specific area and strategies to solve problems. The Fuzzy expert system considers decision-making and problem-solving despite uncertainty and imprecision. In the last decade, the fuzzy expert system has found a vast number of advantages in the medical field as in the diagnosis of child anemia, skin cancer, prostate cancer, brain tumor, breast cancer, cardiovascular diseases, hypertension, ovarian cancer, liver disorder, and more diseases have been diagnosed accurately. Expertise is the principle behind FES, a vast body of detailed knowledge of diagnosing is fed from human to computer [9]. In any system with high ambiguity, vagueness, and complexity, fuzzy logic is appropriate for modelling. When T-1 is compared with T-2 FLS it is found that T-2 FS can regulate the uncertainties efficiently because T-2 FS has more parameters. Figure 11 depicts the components of T-2 FLS.

The process of diagnosing a disease is always uncertain and complicated. The decision about the disease is taken based on expert system knowledge and the patient's condition (Fig. 12).

Generally, a rule-based T-2 fuzzy expert system has three main modules:

(1) **Knowledge base**—This module contains all the rules and data that will be used in diagnosing and decision-making.
(2) **Inference engine**—This module infers inputs based on the rules of the knowledge base. This inference process is based on the adopted logic from the knowledge base.
(3) **Working memory**—This module provides space for the temporary storage of input data and rules that are used in the inference module. The three modules communicate by the user interface to the user, the input is taken from the user and final results are transmitted to the user by the user interface.

Diagnosis of leukaemia

The proposed system for diagnosis of leukaemia using T-2 FL includes two stages.

Stage-1 of the Proposed System

The first phase of the proposed system determines if the blood test is healthy or sick considering the factors introduced for the disease by the expert system. In T-2

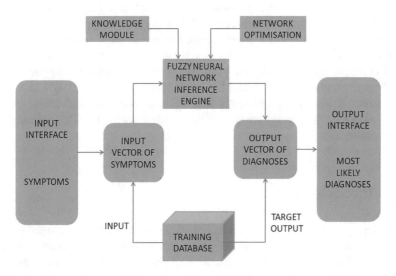

Fig. 11 A schematic diagram of diagnosing disease using fuzzy neural network [17]

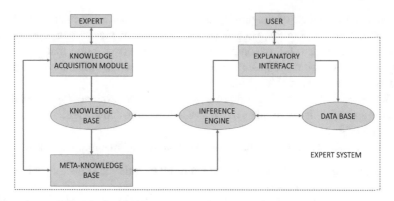

Fig. 12 Components of T-2 FLS [9]

fuzzy modeling of diagnosing leukaemia, platelets, white blood cell, red blood cells, and haemoglobin are taken as system inputs. After that, the system is built using a fuzzy c-means clustering algorithm (Allocating membership functions to each data point matching to each cluster core based on the stretch joining the data point and cluster core) for clustering of the blood test statistics. The proposed index is used for determining new rules. Then, gradually T-1 fuzzy rule base is modified into an interval T-2 rule-based system with unsettled standard deviation and a stable mean. After the rules of the methods are determined, T-2 fuzzy output sets are type-reduced and defuzzified to get the final result [9].

Stage-2 of the Proposed System

In the first stage, we obtain the result of the blood test. In the second stage, symptoms, the period of disease until its improvement, and the output of stage one are combined to obtain the final diagnosis of leukaemia. Following a direct approach for inference, this stage works on expert knowledge. Based on physician knowledge, the marked symptoms will occur in types of leukaemia. If the symptoms suddenly occur, the risk of being diagnosed with ALL and AML will increase, whereas if symptoms occur for a long time, the risk of CLL and CML will increase. The expert system predicts the presence or absence of leukaemia. If a patient is diagnosed with leukaemia, the proposed system determines the type of leukaemia from the knowledge base of the expert system [9].

5 Future Scopes

Fuzzy logic systems have a wide range of scope to be developed especially in the medical field. Intelligent systems and machine learning can be applied to achieve parameters like sensitivity and specificity. T-2 FL can also perform tasks machine learning cannot. Perfection of the models is very important, especially in the fields like the medical field where one cannot afford incorrect classification or false diagnosis [19]. The most common and huge challenge is to make it behave like a diagnostic specialist (like a real human doctor). The accuracy of these models is disputable as they depend on various parameters such as the database used, and the number of factors taken into consideration during patient isolation. When the best algorithm is to be selected while extracting algorithms to their real-time applications, accuracy is not the only parameter to be considered [19]. There has been research continuing on general and interval-2 FLS. One of the future works in T-2 FLS is that the optimization of the footprint of the uncertainty of interval T-2 and general T-2 fuzzy diagnosis system can be fetched by a high-level optimization algorithm. Future study of T-2 has more to do with its design and applications.

6 Conclusions

Fuzzy logic plays an important role in drug industry design, disease management, and medical trials. Fuzzy set theory and all other theories improved from it developed knowledge-based systems and decision-making in medicine. The basic working of T-2 FL and its methods involved in the medical domain brought ease and accuracy to the field. This chapter formalizes different T-2 FL methods concentrating on the medical field and how it depicts its importance in different health monitoring equipment and diagnosing diseases. These systems have formed an integral part of the medical field from diagnosing disease to monitoring surgeries. T-2 fuzzy expert systems have diagnosed leukaemia with an accuracy of 97% which makes it quite reliable. The uncertainty in diagnosis, chiefly in cases where lingual terms are utilized, is run efficiently by T-2 FLS. In the case of high uncertainty in data, GT2 fuzzy logic systems are more accurate compared to IT2 fuzzy logic systems. [20]

References

1. Zohuri, B., Moghaddam, M.: Business Resilience System (BRS): Driven Through Boolean, Fuzzy Logics and Cloud Computation, vol. 11. Springer International Publishing AG (2017)
2. Mendel, J.: Advances in type-2 fuzzy sets and systems. Inform. Sci. **177**, 84–110 (2007)
3. Castillo, O., Melin. P.: Optimization of type-2 fuzzy systems based on bio-inspired methods: a concise review. Inform. Sci. **205**, 1–19 (2012)
4. Karnik, N.N., Mendel, J.M., Liang, Q.: Type-2 fuzzy logic systems. IEEE Trans. Fuzzy Syst. **7**(6), 643–658 (1999)
5. Das, A., Bera, U.K., Maiti, M.: Defuzzification and application of trapezoidal type-2 fuzzy variables to green solid transportation problem. Soft Comput. **22**(7), 2275–2297 (2018)
6. Doostparast, A., Zarandi, M., Zakeri, H.: On type-reduction of type-2 fuzzy sets: a review. Appl. Soft Comput. **27**, 614–627 (2015)

7. Orooji, A., Langarizadeh, M., Hassanzad, M., Zarkesh, M.: A comparison between fuzzy type-1 and type-2 systems in medical decision making: a systematic review. Crescent J. Med. Biol. Sci. **6**(3), 246–252 (2019)
8. Bajpai, A., Kushwah, V.S.: Importance of fuzzy logic and application areas in engineering research. Int. J. Recent Technol. Eng. IJRTE **7**, 1467–1471 (2019)
9. Asl, A.A.S.: A two-stage expert system for diagnosis of leukemia based on type-2 fuzzy logic. Int. J. Comput. Inform. Eng. **13**(2), 34–41 (2019)
10. Melin, P., Castillo, O., Kacprzyk, J., Reformat, M., Melek, W.: Fuzzy Logic in Intelligent System Design: Theory and Applications, vol. 648. Springer (2017)
11. Castillo, O., Melin, P.: Type-2 Fuzzy Logic: Theory and Applications, pp. 5–28. Springer, Berlin, Heidelberg (2007)
12. Neogi, A., Mondal, A.C., Mandal, S.K.: Computational intelligence using type-2 fuzzy logic framework. In: Handbook of Research on Computational Intelligence for Engineering, Science, and Business, pp. 1–29. IGI Global (2013)
13. Ontiveros, E., Melin, P., Castillo, O.: Comparative study of interval type-2 and general type-2 fuzzy systems in medical diagnosis. Inform. Sci. **525**, 37–53 (2020)
14. Mishraand, S., Prakash, M.: Study of fuzzy logic in medical data analytics. Int. J. Pure Appl. Math. **119**(12), 16321–16342 (2018)
15. Maftouni, M., Turksen, I.B., Zarandi, M.H.F., Roshani, F.: Type-2 fuzzy rule-based expert system for Ankylosing spondylitis diagnosis. In: 2015 Annual Conference of the North American Fuzzy Information Processing Society (NAFIPS) held jointly with 2015 5th World Conference on Soft Computing (WConSC), pp. 1–5 (2015)
16. Rahimi Damirchi-Darasi, S., Fazel Zarandi, M., Turksen, I., Izadi, M.: Type-2 fuzzy rule-based expert system for diagnosis of spinal cord disorders. Scientia Iranica, **26**(1), 445–471 (2019)
17. Abiyev, R., Altıparmak, H.: Type-2 Fuzzy neural system for diagnosis of diabetes. Math. Prob. Eng. 1–9 (2021)
18. Vlamou, E, Papadopoulos, B.: Fuzzy logic systems and medical applications. AIMS Neurosci. **6**(4), 266–272 (2019)
19. Thakkar, H., et al.: Comparative anatomization of data mining and fuzzy logic techniques used in diabetes prognosis. Clin. eHealth **4**, 12–23 (2021)
20. Alatrash, M., et al.: Application of type-2 fuzzy logic to healthcare literature search at point of care. In: 2011 Annual Meeting of the North American Fuzzy Information Processing Society. IEEE (2011)

Type-2 Fuzzy Set Approach to Image Analysis

K. Anitha and Debabrata Datta

Abstract Segmentation is an essential task in image analysis process. Due to non-homogeneous intensities, blurred boundaries, noise and minimum of contrast it is a challenging task for image analysists. It has wide range of applications in all fields exclusively in the field medical imaging for disease deionization and early detection. The root cause for non-homogeneous intensities is uncertainty. Various tools have been introduced to handle uncertainty. We have introduced type-2 fuzzy based image segmentation process for edge detection in blurred areas of an image. When compared with classical fuzzy set, it has upper and lower membership values. Since it has more membership values it can handle higher level of uncertainty. In this chapter we have proposed equivalence function associated with strong negation relation which will address each intensity $\mathcal{I}_t$ of an image $\mathcal{I}$ through the membership values. This method is verified with thermographic breast cancer image data set and the results were satisfactory.

Keywords Type-2 fuzzy set · Interval type-2 fuzzy set (IT2FS) · Footprint of uncertainty (FoU) · Fuzzy image

1 Introduction

Image processing is the technique of converting low quality image into high quality image based on certain algorithms for extracting optimal information. There are two major classifications like analog and digital image processing. Analog process handles only hard copies where as digital copies are used in digital image processing techniques. This chapter deals about digital image processing based on type-2 fuzzy set. Classification of digital image, feature selection and extraction from an

K. Anitha (✉)
Department of Mathematics, SRM Institute of Science and Technology-Ramapuram,
Chennai 600089, India
e-mail: anithamaths2019@gmail.com; anithak1@srmist.edu.in

D. Datta
Former Nuclear Scientist, Bhabha Atomic Research Centre, Mumbai 400085, India

© The Author(s), under exclusive license to Springer Nature Switzerland AG 2023
O. Castillo and A. Kumar (eds.), *Recent Trends on Type-2 Fuzzy Logic Systems: Theory,
Methodology and Applications*, Studies in Fuzziness and Soft Computing 425,
https://doi.org/10.1007/978-3-031-26332-3_12

image, multi-scale signal analysis, pattern recognition, hidden Markov models, linear filtering, image editing, image enhancement and image restoration are some of the techniques used in this process. It has the following basic steps:

(i)   **Importing an image**: In this process image can be imported with the help of acquisition tools using single sensor, line sensor and array sensor.

(ii) **Image enhancement**: In this step image is being prepared for further analyzation by doing noise removal, contrast adjustment and sharpening the image.

(iii) **Multi resolution and wavelets**: Here images are converted in to various levels or degrees of resolution which brings time frequency information about the image.

(iv) **Compression**: Image compression helps to reduce the storage space.

(v)   **Segmentation**: This process partitions the digital image into different segments to identify the boundaries and location of object.

At last processed image is coded into the form of information or knowledge base which is elaborated in (Fig. 1).

Medical image processing is one of the most innovative and useful techniques in digital image analysation. In this process computer tomography (CT), Magnetic Resonance Images (MRI), Positron Emission Tomography (PET), Optical Coherence Tomography, Ultrasound, Single Photon Emission Computed Tomography (SPECT), Optical Coherence Tomography (OCT) and Mammographic images are being considered (Fig. 2). It offers non-invasive technique to capture the structural information about a patient or a group of people.

Computed tomography (CT) produced sinogram images which contains detailed information more than the conventional X-ray. Positron Emission Tomography (PET) is the latest method which produces 3D images and it is especially used to study the brain function. MRI is a powerful technique which concentrates on soft tissues and it uses magnetic field to produce radio waves. It identifies the difference between grey and white matter in the brain. Single-Photon Emission Computed Tomography

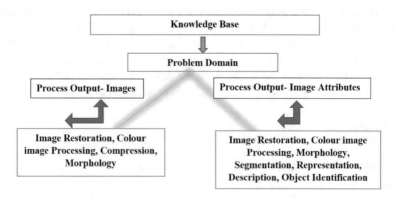

Fig. 1 Image transformation

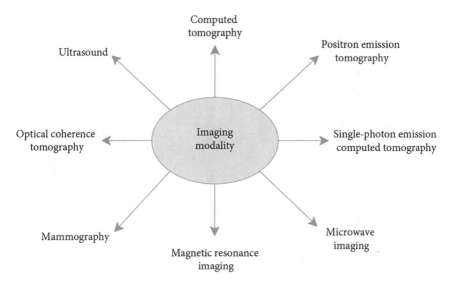

Fig. 2 Image modality

(SPECT) is generally used nuclear image processing through radioactive tracers which analyses neurological disorders. All imaging paradigm takes pivotal role in clinical applications like chemotherapy, radiotherapy, pathology, computer-aided diagnosis and periodic monitoring of patients. Active contour process is one of the image segmentation processes which fixes the boundary of an image with respect to the area of interest. Contour gives smooth shape and closed region for the image which will enable the clinician to analyse the characteristics from CT, MRI, Thermographic images for early detection of diseases.

1.1 Soft Computing Approaches in Image Processing

Soft computing technique is the process of combining computational methods with biological models. The main objective of soft computing techniques used in medical imaging is to handle uncertainties in image data. Fuzzy logic, Genetic algorithm, Rough set, Neural network and Artificial Intelligence are some of the soft computing approaches. Chouhan et al. [5] reviewed various image processing techniques based on soft computing especially fuzzy image processing and segmentation. Figure 3 illustrates various forms of soft computing techniques.

Prof. Zadeh [27] proposed the conceptualization of fuzzy logic in the year 1965. It is a logical approach to handle uncertainty and vagueness with truth values in between 0 and 1. It employs the partial truth through membership values. Gaussian fuzzifier, Singleton fuzzifier are basic types in the process of fuzzification. Rough set theory handles uncertainty through indiscernibility relation which was introduced

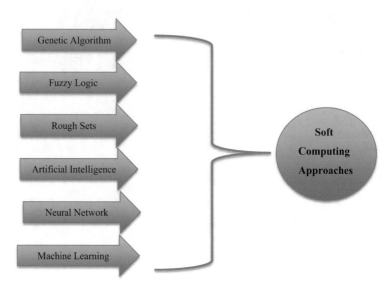

Fig. 3 Soft computing approaches

by Pawlak in 1982. This theory classifies the objects based on boundary values which is the difference between upper and lower approximation. Genetic algorithm is a subset of evolutionary computation. It tries to find the optimal solution from population of possible solution by doing replication process. This method is used in image segmentation and image enhancement with spatial and frequency domain method. But construction of objective function along with its operators is difficult and time consuming. Also, this method fails to handle uncertainty. All above mentioned concepts are lacking in handling uncertainty. Type 2 fuzzy set is the generalized form of type 1 fuzzy set which handles uncertainty more than type 1 fuzzy.

Zadeh [27] constructed fuzzy set with operators in view of variables having linguistic values used to approximate the reasoning. Existence of uncertainty in signals are handled with the help of fuzzy logic [14]. Hagras [10] used fuzzy logic controls in automated robots. Deep insights about type 2-fuzzy sets and systems are discussed in [15, 16]. Ensafi and Tizhooash [8] developed partially dependent approach of image enhancement through Type-2 fuzzy image processor using Type-2 fuzzy upper and lower membership values which are defined as follows:

$$\mu_{UPP}(y) = \mu(y)^{0.5}$$

$$\mu_{LOW}(y) = \mu(y)^2$$

where $\mu(y) \in [0\ 1]$.

Lesion features in coordination of maximum energy and radial distance by trained ANN is introduced by Shan et al. [23]. Claudia proposedType-2 fuzzy logic based edge detection algorithm to handle intrinsic uncertainty [9] along with simulated results. Various soft computing techniques in bio medical imaging are analysed by Devi et al. [7]. Castillo et al. [4] presented the survey about type 2 fuzzy image processing applications in the aspects of image segmentation, edge detection and filtering the image. Krinidis and Chatzis [13] proposed active based contour algorithm to detect an object from an image using curve evolution technique and concluded that fuzzy energy based active contour is more efficient than conventional snake methods. Pereira and Bastos [22] proposed different types of edge detection models. Xian et al. [26] proposed fully automatic breast ultra sound image segmentation process and introduced cost function to find global optimum. Castillo et al. [3] presented literature survey about various image processing techniques like segmenting an image, filtering, classification and edge detection through type-1 fuzzy, type-2 fuzzy and interval valued type 2 fuzzy sets. Mijares et al. [17] used CNN techniques and demonstrated comparative results with other classification techniques for thermogram image analysis. Jafar et al. [12] used RT2FNN network for image interpolation. Murugeswari and Vijayalakshmi [18] introduced interval type-2 CNN method for image classification. A distributed approach type-2 fuzzy logic algorithm was proposed by Benchara [1] and it is implemented in MRI data set. Shikkenawis and Mitra [24] proposed new membership function based on orthogonal basis for improving the image quality of grayscale images. Idris and Ismail [11] proposed fuzzy ID3 algorithm for breast cancer disease classification. They used fuzzy decision tree for disease classification.

Preliminary concepts of type-2 fuzzy sets are demonstrated in Sects. 2 and 3 explains type-2 fuzzy image contouring process, Sect. 4 exhibit the proposed work followed by experimental results and conclusion in Sects. 5 and 6.

2 Type 2 Fuzzy Set- Notation and Terminology

Definition 2.1 (*Fuzzy Set*) A fuzzy subset $\mathcal{F}$ of universe $\mathbb{U}$ is illustrated by its membership function $\mu_{\mathcal{F}} : \mathbb{U} \to [0, 1]$ which associates with each element $u \in \mathbb{U}$, a corresponding number $\mu_{\mathcal{F}}(u) \in [0, 1]$ where $\mu_{\mathcal{F}}(u)$ is graded membership value. The set of all possible points of $\mathbb{U}$ with positive $\mu_{\mathcal{F}}(u)$ values form support of $\mathcal{F}$ and its supremum value is known as height of $\mathcal{F}$.

There are various reasons for existence of uncertainties in type 1 fuzzy sets. When same word gives different meaning in different regions, decision is taken from different people but rate of acceptance may be differed, uncertain and noisy. Model cannot be constructed by type 1 fuzzy for the above uncertain cases. Type-2 fuzzy able to overcome these uncertainty issues since it has three-dimensional membership values. Zadeh introduced the concepts of type-2 fuzzy sets in Zadeh [27].

Definition 2.2 (*Type-2 Fuzzy set*) ($\breve{\mathcal{F}}$) Type-2 fuzzy set corresponds to type-2 membership function $\mu_(\breve{\mathcal{F}})(x, u), x \in X, u \in \mathbb{J}_x \subseteq [0, 1]$. Hence ($\breve{\mathcal{F}}$) is defined by

$$(\breve{\mathcal{F}}) = \left\{ \left((x, u), \mu_(\breve{\mathcal{F}})(x, u)\right) | \forall x \in X, \forall u \in \mathbb{J}_x \subseteq [0, 1] \right\} \tag{1}$$

where three-dimensional membership function $\mu_(\breve{\mathcal{F}})(x, u)$ takes its range $0 \leq \mu_(\breve{\mathcal{F}})(x, u) \leq 1$ and $\mathbb{J}_x$ primary membership of x, $\forall x \in X, x = x'$, $(u, \mu_(\breve{\mathcal{F}})(x', u))$ denotes vertical slice of $\mu_(\breve{\mathcal{F}})(x', u)$ and it is denoted by

$$\mu_{\breve{\mathcal{F}}}(x = x', u) = \int f_{x'}(u)/u \quad, u \in \mathbb{J}_{x'}$$
$$0 \leq f_{x'}(u) \leq 1 \tag{2}$$

Another way of representing ($\breve{\mathscr{F}}$) is

$$(\breve{\mathcal{F}}) = \begin{cases} \int \int \mu_(\breve{\mathcal{F}})(x, u)/(x, u)\mathbb{J}_x \subseteq [0, 1] \to \textit{for continuous case} \\ \sum_{x \in X} \sum_{u \in \mathbb{V}_x} \mu_(\breve{\mathcal{F}})(x, u)/(x, u)\mathbb{J}_x \subseteq [0, 1] \to \textit{for discrete case} \end{cases} \tag{3}$$

Here $\int\int$ *denotes* union of all admissible u and x. The amplitude of secondary membership function is known as secondary grade.

Example Let $X = \{1, 2, 3\}, \mathbb{J}_{x_1} = \{0.2, 0.3, 0.7\}, \mathbb{J}_{x_2} = \{0.4, 1\}, \mathbb{J}_{x_3} = \{0.3, 0.4, 0.8\}$

After assigning grades for elements of $\mathbb{J}_{x_1}$, $\mathbb{J}_{x_2}$ and $\mathbb{J}_{x_3}$ type-2 fuzzy set is defined as follows

$$\breve{\mathcal{F}} = (0.5/0.2 + 1/0.3 + 0.5/0.7) + ((0.5/0.4 + 1/1)/2)$$
$$+ ((0.5/0.3 + 1/0.4 + 0.5/0.8/3)/3)$$

Definition 2.3 (*Interval type-2 fuzzy set (IT2FS)*) In type-2 fuzzy set the variable the secondary variable $u \in [0, 1]$, at each $x \in X, \forall u \in [0, 1]$, $if \mu_(\breve{\mathcal{F}})(x, u) = 1$ then ($\breve{\mathcal{F}}$) is called IT2FS.

The secondary membership function is a vertical slice of $\mu_(\breve{\mathcal{F}})(x, u)$ which is defined by

$$\mu_{\mathcal{F}}(x = x', u) = \int f_x(u)/u \text{ where } u \in \mathbb{J}_x \subseteq [0, 1] \tag{4}$$

Definition 2.4 (*Footprint of uncertainty*) (*FoU*) The essential membership function of ($\breve{\mathcal{F}}$) contains uncertainty. Union of all primary membership forms boundary region of this uncertainty and it is known as footprint of uncertainty (FoU)

$$FoU\left((\breve{\mathcal{F}})\right) = \bigcup_{x \in X} \mathbb{J}_x \tag{5}$$

FoU of IT2FS is defined by.

$$FoU(\mathcal{F}) = \bigcup_{\forall x \in X} \left[\underline{\mu_{\mathcal{F}}(x)}, \overline{\mu_{\mathcal{F}}(x)}\right], \text{ where } \underline{\mu_{\mathcal{F}}(x)}, \overline{\mu_{\mathcal{F}}(x)} \text{ are upper and lower}$$

membership functions associated with upper and lower bounds of $FoU(\mathcal{F})$.

Definition 2.5 (*Fuzzy Image*) Let $\mathcal{I}_{mn}$ is intensity of the image $\mathcal{I}$ in mnth pixel, $m = 1, 2, ..M; n = 1, 2, ... N$. If μ_{mn} its membership value then fuzzy image [8] $\mathcal{F}_{\mathcal{I}}$ is defined as

$$\mathcal{F}_{\mathcal{I}} = \iint_{m,n=1,2...}^{M,N} \frac{\mu_{mn}}{\mathcal{I}_{mn}} \tag{6}$$

Each grey level membership is calculated by

$$\mu(\mathcal{I}_{mn}) = \frac{\mathcal{I}_{mn} - \mathcal{I}_{Min}}{\mathcal{I}_{Max} - \mathcal{I}_{Min}} \tag{7}$$

One of the most commonly used membership functions to fuzzify the image is $\mathcal{S} - function$,

$$\mu(\mathcal{I}(x, y)) = \mathcal{S}(\mathcal{I}(x, y); p, q, r)$$

$$= \begin{cases} 0 & I(x, y) \leq p \\ \frac{[\mathcal{I}(x,y)-p]^2}{(q-p)(r-p)} & p \leq I(x, y) \leq q \\ 1 - \frac{[\mathcal{I}(x,y)-r]^2}{(r-q)(r-a)} & q \leq I(x, y) \leq r \\ 1 & I(x, y) \geq r \end{cases} \tag{8}$$

3 Type-2 Fuzzy on Image Contouring

The essential step in image processing is image segmentation which is still complicated due to the noise, quality and uncertainty of several images. Active contour model is one of the key techniques in image segmentation process. This model moves with a curve or snakes which tracks the restricted characteristics of imposed image to minimize the energy function. Type-2 fuzzy has more degrees of freedom when compared with type1 fuzzy which offers to handle the possibility of more uncertainty. Various types of membership functions like Gaussian, trapezoidal and triangular can be available for both type-2 fuzzy set and IT2FS. The following figures represent membership function of IT2FS and type-2 fuzzy set (Figs. 4 and 5).

Lots of research is going on in the field of image segmentation process with the help of IT2FS and general type-2 fuzzy set techniques. Dawoud [6] proposed threshold method along with type 2 fuzzy measures and tested the effectiveness with thermoscopic images. Nguyen et al. [19] et al. proposed novel genetic IT2 FCM clustering algorithm. In this algorithm chromosome pixels are divided into two clusters which are act as mask for remaining channels and they used M-FISH images for testing their algorithm. Palanivel and Duraisamy proposed fuzzy cluster-based colour texture image segmentation method [20, 21]. In this method they used

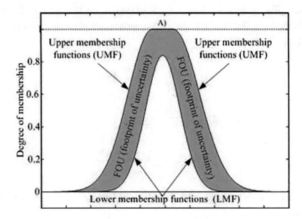

Fig. 4 Interval valued Type-2 fuzzy-Gaussian-membership function

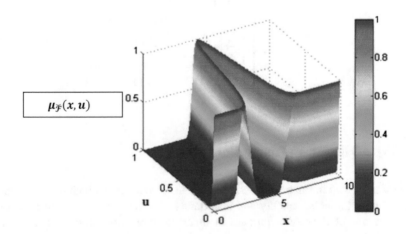

Fig. 5 Type-2 fuzzy-Gaussian-membership function

ICICM method to extract harlick features and exhibited comparative results with other conventional techniques.

Let K be the curve which emerges in the image domain Ω, an image $\mathcal{I}$ compressed in two regions then the energy function is demonstrated as [22]

$$Min E_f(K, k_1, k_2, m) = \mu \, length(K) + \gamma_1 \int [m(x, y)]^r |\mathcal{I}(x, y) - k_1|^2 dxdy$$
$$+ \gamma_2 \int [1 - m(x, y)]^r |\mathcal{I}(x, y) - k_2|^2 dxdy \qquad (9)$$

$$k_1 = \frac{\int [m(x, y)]^r \mathcal{I}(x, y) dxdy}{\int [m(x, y)]^r dxdy} \qquad (10)$$

$$k_2 = \frac{\int [1 - m(x, y)]^r \mathcal{I}(x, y) dxdy}{\int [1 - m(x, y)]^r dxdy} \qquad (11)$$

where m is the membership function, r weight exponent of fuzzy membership, $\mu \geq 0$, fixed values $k_1, k_2 > 0$ then the membership of each pixel is calculated by

$$m(x, y) = \frac{1}{1 + \left(\frac{\gamma_1 (\mathcal{I}(x,y) - k_1)^2}{\gamma_2 (\mathcal{I}(x,y) - k_2)^2}\right)^{\frac{1}{r-1}}} \qquad (12)$$

$$Length \, K = \sum_{i,j} \sqrt{\left(P_{i+1,j} - P_{i,j}\right)^2 + \left(P_{i,j+1} - P_{i,j}\right)^2} \qquad (13)$$

where $P_{i,j} = H[m(x, y) - 0.5]$, H—Heavyside function.

Various optimization techniques have been proposed to solve this objective function.

The inner, outer and curve region of the image is defined as

$$\mathbb{S}_{\mathcal{L}} = \begin{cases} K \Rightarrow \{(x, y) \in \mathcal{I} | m(x, y) = 0.5\} \\ inside(K) \Rightarrow \{(x, y) \in \mathcal{I} | m(x, y) > 0.5\} \\ outside(K) \Rightarrow \{(x, y) \in \mathcal{I} | m(x, y) < 0.5\} \end{cases} \qquad (14)$$

4 Proposed Method by Type-2 Fuzzy Set

Let us defined the equivalence function $\varphi : [0, 1]^2 \rightarrow [0, 1]$ with associated with strong negation relation$\backslash$. If $\theta_1(x) = x, \theta_2(y) = y$ be automorphisms then

$$\varphi(x, y) = 1 - |x - y| \text{ with strong negation } n(x) = 1 - x \qquad (15)$$

For type-2 fuzzy set the corresponding equivalence function is defined as
$\zeta : [0, 1] \rightarrow [0.5, 1]$ such that

$$\begin{cases} \zeta(x) = 1 & if \quad x = 1 \\ \zeta(x) = 0.5 & if \quad f \ 0 < x < 1 \\ \zeta(x) \ is \ an \ increasing \ function \end{cases}$$

Then for an image $\mathcal{I}$, and each intensity of the set $\mathcal{I}_t$ the membership function is given by

$$\mu_{\mathcal{I}_t}(a) = \begin{cases} \zeta[\varphi(a, \ d_b(t))], & if \quad a \leq t \\ \zeta[\varphi(a, \ d_0(t))], & if \quad a > t \end{cases} \tag{16}$$

$$d_b(t) = \frac{\sum_{a=0}^{t} an(a)}{\sum_{a=0}^{t} n(a)}; \ d_0(t) = \frac{\sum_{a=t+1}^{T-1} an(a)}{\sum_{a=t+1}^{T-1} n(a)} \tag{17}$$

where $n(a)$—number of pixels with intensity a

$T[0, 1] \subseteq [0, 1]$, T—Set of all closed sub-intervals.

Now the corresponding S function (8) will be converted to

$$S(a, \ p, q, r) = \begin{cases} 0 & a \leq p \\ 2\left(\frac{a-p}{a-r}\right)^2 & p \leq a \leq q \\ 1 - \ 2\left(\frac{a-r}{r-p}\right)^2 & q \leq a \leq r \\ 1 & a \geq r \end{cases} \tag{18}$$

The proposed method has following steps.

(i)   Import the image
(ii) Image contouring by (9)
(iii) Fuzzification of contoured image using $S(a, p, q, r)$ in (18)
(iv) Evaluation of segmented result.

Evaluation process needs true positive (TP), similarity (SI) and false positive (FP). Let.

S_A be the objection region selected by proposed method, S_R-real region of object then error metrices are given by

$$\begin{cases} TP = \frac{|S_A \cap S_R|}{|S_R|} \\ FP = \frac{|S_A \cup S_R - S_R|}{|S_R|} \\ SI = \frac{|S_A \cap S_R|}{|S_A \cup S_R|} \end{cases} \tag{19}$$

5   Experimental Results

The data set provided by Silva et al. [25] and it can be downloaded from [2]. This set contains 63 thermographic images in RGB, JPEG format with dimension 680 × 480 ×3. Our proposed method give efficient result for different thermographic image and results are displayed as follows. Image-j software is used for contouring process and fuzzification process is taken place by MATLAB 2021b (Figs. 6, 7, 8, 9, 10 and 11).

Segmentation process done through boundary and region level. Goodness measure is used to evaluate the performance and the results are displayed as follows.

Average True Positive Value: 99.24%

Average False Positive Value: 19.35%

Similarity Index: 86.7

Accuracy:
$$\frac{True Positive + True Negative}{True Positive + True Negative + False Positive + False Positive}$$
$$= 98.23\%$$

Fig. 6 Input image-original thermographic image

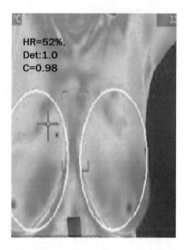

Fig. 7 Gaussian blur 3D filter

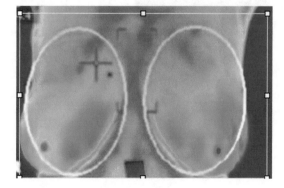

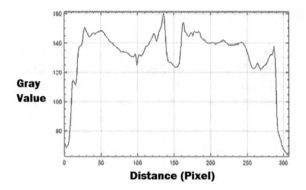

Fig. 8 Grey level graph

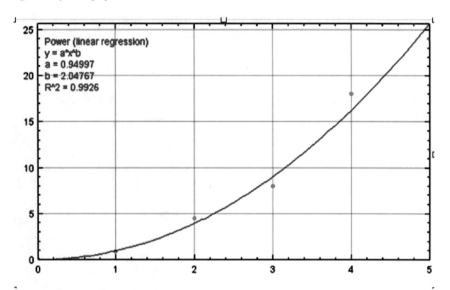

Fig. 9 Linear Regression graph of Filtered image

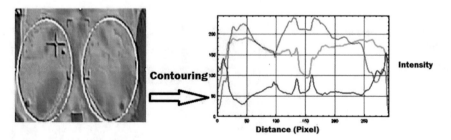

Fig. 10 Image contouring with RGB graph

Fig. 11 Output of
fuzzification through type-2
fuzzy sets

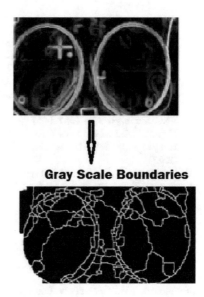

This method gives maximum accuracy in the process of segmentation in both boundary and region level.

6 Conclusion

We exhibited the mathematical concepts about type-2 fuzzy sets along with its properties. Applications of type-2 fuzzy set in image processing and its role in image segmentation, edge detection are briefly discussed. Since Type-2 fuzzy set covers maximum level of uncertainty it can handle uncertainty in various fields of decision making. Experimental results shows that the proposed method give optimal result in various forms image analysis process.

References

1. Benchara, F.Z., Youssfi, M.: A new distributed type-2 fuzzy logic method for efficient data science models of medical informatics. Adv. Fuzzy Syst. (2020). https://doi.org/10.1155/2020/6539123
2. Bottema, M.J., Slavotinek, J.P.: Detection and classification of lobular and DCIs microcalcifications in digital mammograms. Pattern Recogn. Lett. **24**, 1209–1214 (2000)
3. Castillo, O., Sanchez, M.A., Gonzalez, C.I., Martinez, G.E.: Review of recent Type-2 fuzzy image processing applications. Information **8**(97), 1–18 (2017)
4. Castillo, O., Melin, P., Castro, J.R.: Computational Intelligence Software for Interval Type-2 Fuzzy Logic. Proceedings of the 2008 Workshop on Building Computational Intelligence and Machine Learning Virtual Organizations, pp. 9–29 (2008)

5. Chouhan, S.S., Kaul, A., Singh, U.P.: Soft computing approaches for image segmentation-a survey. Multimed Tools Appl. **77**, 28483–28537 (2018)
6. Dawoud, A.: Segmentation of Dermoscopic images by the fusion of Type-2 fuzziness measure in graph cuts image Binarization. Int. J. Imaging Robot **2**(15), 73–87 (2015)
7. Devi, M., Singh, S., Tiwari, S., Patel, S.C., Ayana, M.T.: A survey of soft computing approaches in biomedical imaging. Hindawi J. Healthcare Eng. **21**, 2040–2295 (2021)
8. Ensafi, P., H.R., Tizhoosh, R.: Type-2 Fuzzy Image Enhancement. Springer, Berlin, Heidelberg, LNCS 3656, pp. 159–166 (2005)
9. Gonzalez, C.I., Melin, P., Castillo, O.: Edge detection method based on general Type-2 fuzzy logic applied to colour images. Information **8**(104), https://doi.org/10.3390/info8030104 (2017)
10. Hagras, H.A.: A Hierarchical Type-2 fuzzy logic control architecture for autonomous mobile robots. IEEE Trans. Fuzzy Syst. **12**, 524–539 (2004)
11. Idris, N.F., Ismail, M.A.: Breast cancer disease classification using fuzzy-ID3 algorithm with FUZZYDBD method: automatic fuzzy database definition. Peer J. Comput. Sci., e427 (2021). https://doi.org/10.7717/peerj-cs.427
12. Jafar, T., Chunwei, Z., Ardashir, M.Z., Saleh, M., Mosavi Amir, H.: Medical image interpolation using recurrent Type-2 Fuzzy neural network. Front. Neuro-Inf. **15**(75) (2021)
13. Krinidis, S., Chatzis, V.: Fuzzy Energy-based active contours. Trans. Image Process. **18**, 2747–2755 (2009)
14. Mendel, J.M.: Uncertainty, fuzzy logic, and signal processing. Signal Process **80**, 913–933 (2000)
15. Mendel, J.M.: Type-2 Fuzzy Sets and Systems—A Retrospective, pp. 523–532. Springer, Berlin, Heidelberg. https://doi.org/10.1007/s00287-015-0927-4 (2015)
16. Mendel, J.M., Bob John, R.I.: Type-2 fuzzy sets made simple. IEEE Trans. Fuzzy Syst. **10**(2), 117–127 (2002)
17. Mijares, S.T., Woo, F., Flores, F.: Breast cancer identification via thermography image segmentation with a gradient vector flow and a convolutional neural network. J. Healthcare Eng. **98**(19). https://doi.org/10.1155/2019/980619 (2019)
18. Murugeswari, P., Vijayalakshmi, S.: New method of Interval type-2 fuzzy-based CNN for image classification. Int. J. Fuzzy Logic Intell. Syst. **20**, 336–345 (2020)
19. Nguyen, D.D., Ngo, L.T., Watada, J.: A genetic type-2 fuzzy C-means clustering approach to M-FISH segmentation. J. Intell. Fuzzy Syst. **27**, 3111–3122 (2014)
20. Palanivel, M., Duraisamy, M.: Adaptive color texture image segmentation using α-cut implemented interval Type-2 Fuzzy C-Means. Res. J. Appl. Sci. **7**, 258–265 (2012)
21. Palanivel, M., Duraisamy, M.: Color textured image segmentation using ICICM-Interval Type-2 Fuzzy C-Means clustering hybrid approach. Eng. J. **16**(5), 115–126 (2012)
22. Pereira, C.L., Bastos, C.A.C.M.: Tsang Ing Ren & George D.C. Cavalcanti: fuzzy active contour models. IEEE International Conference on Fuzzy Systems (FUZZ-IEEE 2011), pp. 1621–1627 (2011)
23. Shan, J., Cheng, H.D., Wang, Y.: Completely automated segmentation approach for breast ultrasound images using multiple-domain features. Ultrasound Med. Biol. **38**(2), 262–275 (2012)
24. Shikkenawis, G., Mitra, S.K.: Image denoising using 2D orthogonal locality preserving discriminant projection. J. IET Image Process. **14**(3), 554–560 (2019)
25. Silva, L.F., Saade, D.C.M., Sequeiros, G.O.: A new database for breast research with infrared image. J. Med. Imaging Health Inf. **4**(1), 92–100 (2014)
26. Xian, M., Zhang, Y., Cheng, H.D.: Fully automatic segmentation of breast ultrasound images based on breast characteristics in space and frequency domains. Pattern Recogn. **48**(2), 485–497 (2015)
27. Zadeh, L.A.: Fuzzy sets. Inf. Control **8**(3), 338–353 (1965)

Non-iterative Wagner-Hagras General Type-2 Mamdani Singleton Fuzzy Logic System Optimized by Central Composite Design in Quality Assurance by Image Processing

Pascual Noradino Montes Dorantes and Gerardo Maximiliano Mendez

Abstract This paper presents the implementation of a non-iterative General Type-2 (GT2) Fuzzy Logic System (FLS) in quality assurance by image processing using Mamdani singleton model based on Wagner-Hagras (WH) algorithm. The antecedents and consequents are modelled and remain fixed. The modelling of the rule base uses the Central Composite Design (CCD) model to create a classifier in an industrial quality area. Results show that the implementation of the WH GT2FLS model provides very close or better results with a few alpha-cut versus an Interval Type-2 model (IT-2) FLS system depending on the type of membership function selected for the system.

Keywords General type-2 · GT2 FLS · Fuzzy logic · Quality assurance · Artificial vision quality assurance · Non-contact measurement

1 Introduction

1.1 GT-2 Literature

The GT2 models show slow development due to various challenges as mentioned by different authors [1–15]. The existing literature presents theoretical or experimental proposals and as is mentioned by Mendel in [1] the GT2 technology is in their infancy. The challenges that must be faced are listed in (Table 1) and due to these challenges,

P. N. Montes Dorantes (✉)
División de Estudios de Posgrado e Investigación, Universidad Autónoma del Noreste, Blvd. José Musa de León y General Medardo de La Peña S/N, Col. Los Pinos, Saltillo, Coahuila 25100, México
e-mail: pascualresearch@gmail.com

G. M. Mendez
Departamento de Ingeniería Eléctrica y Electrónica, Instituto Tecnológico de Nuevo León, Cd. Guadalupe N. L., Guadalupe 67170, México
e-mail: gerardo.m@nuevoleon.tecnm.mx

© The Author(s), under exclusive license to Springer Nature Switzerland AG 2023
O. Castillo and A. Kumar (eds.), *Recent Trends on Type-2 Fuzzy Logic Systems: Theory, Methodology and Applications*, Studies in Fuzziness and Soft Computing 425,
https://doi.org/10.1007/978-3-031-26332-3_13

Table 1 Implementation challenges of GT2 model

Challenge	References
Difficult to implement	[2]
Information are non-functional	[3]
Information is un useful	[3]
Information not needed	[3]
Complex learning process	[5–9]
Hard computation	[5, 8, 9, 11–13]
Defuzzification very complex	[5, 13, 14]
Exhaustive computational time	[5, 8, 9, 11–13]
Impractical to usage	[5]
Method iterative and algorithmic	[15]
Determination of the number of alpha planes	[11]

the development of GT2 applications is not widely documented in the literature as corroborated by [7, 15, 16].

Mendel in [1] recommends the use of optimization models to generate the necessary parameters iteratively until an acceptable performance of the system is obtained. In contrast, several authors [5, 8, 9, 11–13, 17] mention that the GT2 models do not require to be tuned because, they do not reflect the uncertainties of the real world.

On one hand, Melin and Castillo [2] mention that GT2 systems are too much difficult to implement and difficult to use. On the other hand, Gilan et al. [3] and Salehi et al. [5] mentions that the third dimension information is unnecessary, useless, impractical, and non-functional.

The GT2 systems, have a complex learning process as mentioned by [5–9]; their computationally time is expensive [5, 8, 9, 11–13]. The GT2 system defuzzification is very complex [4, 13, 14] and the GT2 system is iterative [15]. Due to the challenges mentioned in previous paragraphs, the applications of the GT2 systems are mostly theoretical or experimental as is shown in Table 2.

To detect and group repetitive information the GT2 model is used in [4], a classification system is presented in [5], while a model to regulate glucose levels is presented in [7]. A diagnosis of depression is generated by GT2 model in [8]. To approximate unknown nonlinear complex functions [9] uses the GT2 model, in [11] is established an experimental application to compare the cost of the implementation of the system based on the number of alpha planes used on GT2 system. A classifier to medical issues is developed in [12], to generate the inspection a GT2 hybrid model to clustering is presented in [14], in [15] is used the GT2 to regulate the traffic in a crossroad.

Classification for medical diagnosis is presented in [16], while [17] uses the GT2 to manage domestic appliances. In [18] is presented an application to regulate the voltage and to improve and to handle disturbances in a power system via GT2 model, in [19] is presented a model to regulate the frequency in micro-grids. Pattern recognition and clustering are generated in [20] by c-means algorithm. A model

Table 2 Brief Survey of GT2 applications in the literature

Authors	Type		
	Theoretical	Experimental	Application
[18]	X	X	Power systems
[19]	X	X	AC micro grids
[20]	X	X	Clustering
[21]		X	Glucose level regulation
[22]		X	Traffic scheduling
[23]		X	Medical diagnosis
[5]	X	X	Classification
[4]	X	X	Clustering
[11]	X	X	Comparing
[24]	X	X	Classifier
[25]	X	X	Inspection
[26]	X	X	Classification
[27]	X	X	Control, Medical issue
[9]	X	X	Regulation
[28]	X	X	Classification
[17]	X	X	Domestic appliances
[14]	X	X	Clustering
[29]	X	X	Medical diagnosis
[30]	X	X	Tuning

to regulate the anesthesia is generated in [27], in [28] is used the GT2 to classify customers. Medical diagnosis is proposed in [29] using GT2, finally in [30] GT2 is used for tuning the system.

Various optimization models have been used to try to implement and optimize GT2 systems such as: Ordered Weighted Averaging (OWA) in [5], [6] uses data-driven, Kalman filters [8], Artificial Neural Networks (ANN) in [9], Least Square Estimator (LSE) [10], in [14] is used a hybrid ANN to optimize clustering, Recursive Least Squares (RLS) [14, 30], Least Squares (LS) in [16], Teaching Learning Based Optimization (TLBO) in [18], a hybrid differential evolution algorithm, Biogeography-Based Optimization (BBO) [21], searching algorithms [31], Particle Swarm Optimization (PSO) in [14, 26, 32], Gradient-based method [32].

The main contribution of this work is the implementation of GT2 model in real world process.

1.2 Problems on Image Acquisition and Processing

Due to the problems generated in the acquisition and processing of digital images such as: Variations produced by the camera, the shape of the lens of the camera which produces variations on image [33]provides an online application. Using this application could be obtained an sketch that shows the defects in the lens and variations in the form of them.

Variations in the image as: The spatial position of the capture device, this position causes distortions on the size of the sample that appears in the picture, this phenomenon is called parallax [34] and cause changes in the dimension of the sample.

By the geometry of the object, the parallax changes the spatial positioning of the object in a 3D environment [35] and produces perspective that derives on changes of shape and dimension. The position of an edge of the object at the border of a pixel causes loss of information or addition of it. e.g. Demant et al. [36] mentions, that is necessary at least a width of 570 pixels to identify a bar code to obtain at least five pixels for every thin line due to degradation in the filtering and segmentation phases but, only with this condition, the identification is accomplished. The conformation of the sensor, e.g. A charged couple device having several constraints as: The sensor the field of view is very close to the 570 pixels, the loss of information in data transmission is 10%, approximately. With this condition only have 523 pixels and the identification of a bar code cannot be made as mention [36].

The fact of the pixel form are non-square; the thermal excitation of the neighbor pixels in the sensor due to the charge of the photon. Other factors are, the exposition time regulated by the F number [37]. Exists some perturbations caused by the environmental conditions that produce a variation in the measurement data in an amount near to a one standard deviation as is mentioned by the National Institute of Standards and Technology (NIST), [38]. These conditions produce corrupted data and their values are increased or decreased due to the displacement of the Membership Function (MF's) in the FLS to the left or to the right due to the uncertainty [39].

2 Materials and Methods

2.1 Type-2 Fuzzy System

A type-2 fuzzy model also called IT-2 is based in a couple of Type-1 Fuzzy Sets (T1 FS) depicted on (Fig. 1) due to the uncertainty the IT-2 is delimited in an interval delimited by two T1 FS (Fig. 2) [1].

A type-2 fuzzy set is represented by $\tilde{A}$ and presented in (1) that shows the union of all membership functions $\mu_{\tilde{A}}(x, u)$ in a continuous universe restricted to (2).

$$\tilde{A} = \left\{ (x, u), \left(\mu_{\tilde{A}}(x, u) \right) | \forall x \in X, \forall u \in J_x \subseteq [0, 1] \right\} \qquad (1)$$

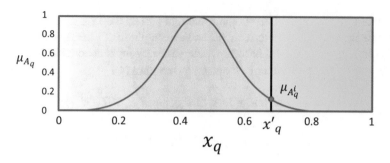

Fig. 1 Type-1 fuzzy set

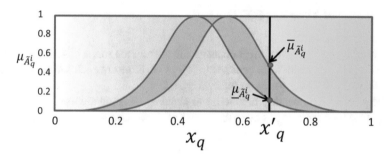

Fig. 2 Type-2 fuzzy set

$$0 \le \mu_{\tilde{A}}(x, u) \le 1 \tag{2}$$

Every value of x'_q shown in Fig. 2 is represented in a plane on (z, y) axis with upper and lower limits. Those limits are called secondary membership function depicted in Fig. 3. The upper and lower limits are commonly called in literature left and right membership grades.

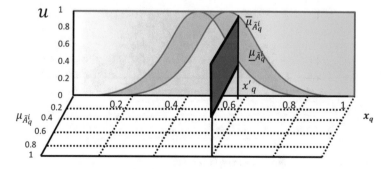

Fig. 3 Secondary membership function on IT-2 model

Equal to the Type-1 Fuzzy Logic System (T1 FLS) the IT-2 FLS depends on a rule base in the form (3) with x_n inputs and one output for T1 SFLS in the IT-2 FLS model the rule changes and have two values at every input denoted by $\tilde{A}_n^i$ and two values at every output given by $\tilde{G}^i$ and taken the form (4).

$$R^i : IF\ x_1\ is\ A_1^i\ and\ x_2 is\ A_2^i\ and\ \dots\ and\ A_n^i\ THEN\ y\ is\ G^i \tag{3}$$

$$\tilde{R}^i : IF\ x_1\ is\ \tilde{A}_1^i\ and\ x_{p=2}\ is\ \tilde{A}_2^i\ THEN\ y\ is\ \tilde{G}^i \tag{4}$$

2.2 General Type -2 (GT-2)

For this case, the secondary MF's changes and acquire the shape of a T1 FS (Fig. 1). The GT-2 secondary MF's are depicted in (Fig. 4) in there are enclosed all alpha-cuts or alpha planes that conforms an Interval type-2.

The primary membership functions are given by (5). And are used the Center of Sets (COS) type reducer to obtain the values of the consequents named: c_l^i and c_r^i,

$$\mu_a(x) = exp\left[-\frac{1}{2}\left[\frac{x_i - \bar{x}_i}{\sigma_{xi}}\right]^2\right] \tag{5}$$

where: $\mu_a(x)$ is the membership function value of the variable x_i, $\bar{x}_i$ is the mean of the fuzzy set and σ_{xi} is the dispersion of the fuzzy set.

The alpha planes are called $\tilde{A}_\alpha$ whose grades are at least equal to alpha or grater and $\alpha \in [0, 1]$. The depiction of an alpha plane is presented in (Fig. 5). Then, the fuzzy set GT-2 type is given by (6) and denoted by $\tilde{A}_\alpha$,

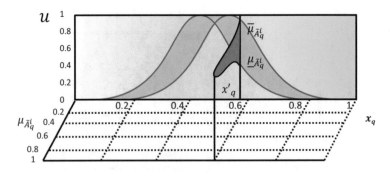

Fig. 4 Secondary membership function on GT2 model

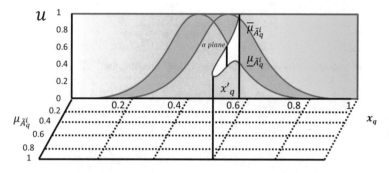

Fig. 5 Alpha plane (y, z axis) secondary membership function of GT2 model

$$\tilde{A}_\alpha = \{(x, u), \mu_{\tilde{A}}(x, u) \geq \alpha | x \in X, u \in [0, 1]\} \tag{6}$$

where:

$$a_\alpha(x) = Lower\ Membership\ Function\ (\tilde{A}_\alpha) \tag{7}$$

$$b_\alpha(x) = Upper\ Membership\ Function\ (\tilde{A}_\alpha) \tag{8}$$

The alpha plane, alpha cut or horizontal slice in the axis (y, z) is defined by (9) and depicted in (Fig. 6).

$$R_{\tilde{A}_\alpha} = \alpha / \tilde{A}_\alpha \tag{9}$$

In the IT-2 models is required a type reduction that converts the type-2 output to type-1 output obtaining a crisp output instead of an interval. In this proposal, the type-reduction is performed by the COS method (9) as is mentioned by Mendel in [1] in the Wagner-Hagras method.

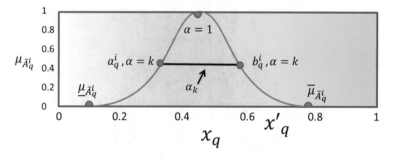

Fig. 6 Alpha plane or horizontal slice on y, z-axis of secondary membership function of GT2 model

$$Y_{COS} = \frac{1}{[y_l^{cos}(x'), y_r^{cos}(x')]} \tag{10}$$

And the upper and lower membership values are obtained by (11, 12).

$$y_l^{cos}(x') = \frac{\sum_{i=1}^{L} c_l(\tilde{G}^i)\overline{f}^i(x') + \sum_{i=1}^{L} c_l(\tilde{G}^i)\underline{f}^i(x')}{\sum_{i=1}^{L} \overline{f}^i(x') + \sum_{i=1}^{L} \underline{f}^i(x')} \tag{11}$$

$$y_r^{cos}(x') = \frac{\sum_{i=1}^{R} c_r(\tilde{G}^i)\overline{f}^i(x') + \sum_{i=1}^{R} c_r(\tilde{G}^i)\underline{f}^i(x')}{\sum_{i=1}^{R} \overline{f}^i(x') + \sum_{i=1}^{R} \underline{f}^i(x')} \tag{12}$$

2.3 Interval Type-2 Singleton Fuzzy Logic System (IT2 SFLS)/CCD Model

The IT2 SFLS/CCD requires a correlation matrix (13) of the input variable (x_i), the additional parameter σ_{x_i} is calculated the IT2 SFLS rule base [40],

$$N^k \tag{13}$$

where: N represents the variables and k represents the possible states of every variable in the control limits.

The upper and lower limits are produced by arithmetical calculus as is presented in (14) and (15), their consequents are given by (16) and (17). Then the matrix to conform rule base for the IT-2 is represented by (18).

$$\underline{N} = x_i - \sigma_{x_i} \tag{14}$$

$$\overline{N} = x_i + \sigma_{x_i} \tag{15}$$

$$\underline{y} = y_i - \sigma_{y_i} \tag{16}$$

$$\overline{y} = y_i + \sigma_{y_i} \tag{17}$$

$$Rb = \begin{bmatrix} \underline{a} & \overline{a} & \underline{b} & \overline{b} & y_1 & \overline{y_1} \\ \underline{a} & \overline{a} & \underline{B} & \overline{B} & y_b & \overline{y_b} \\ \underline{A} & \overline{A} & \underline{b} & \overline{b} & y_a & \overline{y_a} \\ \underline{A} & \overline{A} & \underline{B} & \overline{B} & \underline{y_{ab}} & \overline{y_{ab}} \end{bmatrix} \tag{18}$$

For i = 1, a, b, ab.

Where: Rb represents the fuzzy rule base matrix, $\underline{a}$ represents the lower limit of the IT-2 for the variable a, the upper limit of the IT-2 for the variable a is obtained by $\overline{a}$ in the LCL_a. $\underline{A}$ represents the lower limit of the IT-2 for the variable a, the upper limit for the IT-2 for the variable a is obtained by $\overline{A}$ in the UCL_a, $\underline{b}$ represents the lower limit of the IT-2 for the variable b, the upper limit of the IT2 for the variable b is obtained by $\overline{b}$ in the LCL_b, $\underline{B}$ represents the lower limit of the IT-2 for the variable b, the upper limit of the IT-2 for the variable b is obtained by $\overline{B}$ in the UCL_b. The interval for the output response is obtained by $\underline{Y}_i$, $\overline{Y}_i$. Those values represent the lower and upper output or response for a specific pair of states *(a, b, ab, 1)* or treatments of the inputs.

The Eqs. (14–17), presents the basis to obtain the antecedents of the rule base for GT-2 mode given by (19–22),

$$m_1 = \underline{a} = a - \sigma_a \tag{19}$$

$$m_2 = \overline{a} = a + \sigma_a \tag{20}$$

$$\underline{y}_i = y - \sigma_y \tag{21}$$

$$\overline{y}_i = y + \sigma_y \tag{22}$$

where: m_1 represents the lowest value function, m_2 represents the upper value function, $\underline{a}$ represents the lower limit of the variable a, the upper limit of a is obtained by $\overline{a}$, $\underline{b}$ represents the lower limit of the variable b, the upper limit of b is obtained by $\overline{b}$, the interval for the output response is obtained by $\underline{Y}_1$, $\overline{Y}_1$. Those values represent the lower and upper output or response for a specific pair of states or treatments of the inputs.

3 Theory and Calculation

3.1 Proposal

The model of GT2 system is so similar to the IT-2 model, then; a type 2 rule base could be used to develop a GT2 system. The model presented in [40] serves as a basis to implement a more flexible model as the WH algorithm [1]. To create a GT2 model is required a combination of the data presented in [41] and presented on Table 3 with there, are obtained the values for the rule base (Table 4) in the type-1 singleton model. Using (14–22) are obtained an IT-2 rule base to generate the antecedents and consequents for the GT2 WH model (Tables 5, 6 and 7).

Table 3 Data mining from industrial process image database used for CCD

Samples	X_1	X_2	Goal
1	208	139	0.6
2	214	141	3.6
3	218	135	1.2
4	218	142	5.4
5	210	142	3
6	209	141	2.1
7	205	137	−1.5
8	218	143	6
9	206	141	1.2
10	211	146	5.7
11	211	147	6.3
12	208	143	3
13	207	145	3.9

Table 4 Rule base for Factorial design/2^k/ CCD model

Parameter	Left lower limits	Right lower limits	Left upper limits	Right upper limits
X_2	198	218	198	218
Y	−6	0	0	6

Table 5 Antecedents of first variable, x_1

	L	R	σ
1	130	136	3.3
2	140	146	3.3

Table 6 Antecedents of second variable, x_2

	L	R	σ
1	193	213	6.6
2	203	223	6.6

Table 7 Consequents for four rules

Rule	c_l	c_r
1	−8	−4
2	−2	2
3	−2	2
4	4	8

Assemble of the proposed systems consists in a singleton Gaussian fuzzifier (5), product t-norm (23) from [1] and COS type reduction (10) to final output and (11, 12) to generate interval output left and right. Also use four alpha planes to emulate the [1] example,

$$\mu_{A \cap B}(x) = [\mu_A(x) * \mu_B(x)] \tag{23}$$

where: $\mu_A(x)$, $\mu_B(x)$ represents the membership of Xi respectively at their fuzzy set Xi, T represents any t-norm, both $\mu_A(x)$, $\mu_B(x)$ yields (5).

4 Results

Are used 10 pairs to test the GT2 WH model, the targets are obtained from Table 3. The results obtained for one sample with a non-trained model are shown in Table 8. The Root Mean Square Error (RMSE) was used as indicator to show the precision of the GT2 system with different number of alpha cuts to see their performance in different scenarios (Table 9). In Fig. 7 can be seen that the triangular membership functions shown better values contrasted with Gaussian MF's and the best results are obtained with more than one alpha cut. In contrast, when the results are placed in millimeters, the best results are obtained using the Gaussian MF's and are better in a value near to 25% (Table 10 and Fig. 8).

Table 8 Estimated output, non-trained model. One input–output data pair

	0 α-cut = IT2 = TISFLS	4 α-cuts	10 α-cut
T1 SFLS Gaussian	0.66127478	–	–
IT2 Gaussian	0.441	–	–
WH COS Gaussian		0.4660502	0.460
T1 SFLS Triangular	0.6		
IT2 Triangular	0.675		
WH COS Triangular		0.272	0.130

Table 9 RMSE in prediction (in sigmas)

	0 α-cut = IT2 = TISFLS	GT2 1 α-cuts	GT2 4 α-cuts	GT2 10 α-cut
IT2 Gaussian	1.12			
WH COS Gaussian		1.375	1.33	1.42
IT2 Triangular	0.878			
WH COS Triangular		1.05	0.803	1.129

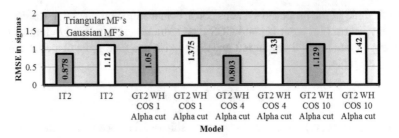

Fig. 7 RMSE of prediction evaluation in sigmas

Table 10 RMSE in prediction (in millimeters)

	0 α-cut = IT2 = TISFLS	GT2 1 α-cuts	GT2 4 α-cuts	GT2 10 α-cut
IT2 Gaussian	1.68			
WH COS Gaussian		2.39	1.99	2.13
IT2 Triangular	2.63			
WH COS Triangular		3.15	2.4	3.38

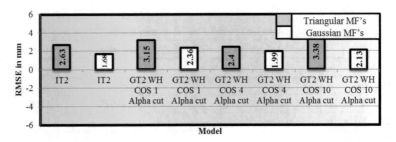

Fig. 8 RMSE of prediction evaluation in millimeters

On Fig. 9 are shown the tolerance zone for the industrial process delimited in ±5 mm demonstrating that the non-contact quality assurance process based on artificial vision is an excellent option to evaluate online the quality of the goods.

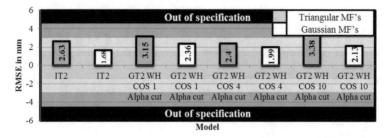

Fig. 9 Quality control process chart. Tolerance zone defined by ±3σ and RMSE of prediction evaluation in millimeters

5 Discussion

As is mentioned in several papers, was proved that in some applications the precision of the model does not depend directly on the training phase. The precision depends on the mayor part in the modeling phase [40–44].

Actually, in all process exists a tolerance as is mentioned by international organizations as International Standard Organization (ISO) [45], NIST [46] and the International Bureau of Weights and pounds (BIMP) [45]. These tolerances are commonly used in the control charts for statistical process control.

The results shown in the previous section provide accurate values without worrying the problems shown in Sect. 1.2 that are generated in the acquisition and processing phase of the images.

All values are placed into the two sigma of specification. That means that the measurement accomplishes the accepted product criteria and the method tested works with accuracy.

6 Conclusions

The most remarkable fact, in this application is that it can be used online to test and evaluate quality features in a production process with an acceptable confidence. In Fig. 9 where can see that the obtained predictions of quality features are in the tolerance zone marked in the industrial control process delimited by ±sigma.

The error obtained in the measurements are below 1.5 sigma and is remarkable that the images used for the test has to low resolution (640 × 480 pixels), with high resolution the error are reduced and this is part of the future work. That means that these classes of systems are an excellent opportunity to evaluate online quality features.

References

1. Mendel, J.M.: Uncertain Rule-Based Fuzzy Systems. Springer, Introduction and new directions (2017)
2. Melin, P., Castillo, O.: An intelligent hybrid approach for industrial quality control combining neural networks, fuzzy logic and fractal theory. Inf. Sci. **177**, 1543–1557 (2007)
3. Gilan, S.S., Sebt, M.H., Shahhosseini, V.: Computing with words for hierarchical competency based selection of personnel in construction companies. Appl. Soft Comput. **12**, 860–871 (2012)
4. Salehi, F., Keyvanpour, M.R., Sharifi, A.: GT2-CFC: general type-2 collaborative fuzzy clustering method. Inf. Sci. **578**, 297–322 (2021)
5. Shahparast, H., Mansoori, E.G.: Developing an online general type-2 fuzzy classifier using evolving type-1 rules. Int. J. Approximate Reasoning **113**, 336–353 (2019)
6. Cheng-Dong, L.I., Gui-Qing, Z.H.A.N.G., Hui-Dong, W.A.N.G., Wei-Na, R.E.N.: Properties and data-driven design of perceptual reasoning method based linguistic dynamic systems. Acta Automatica Sinica. **40**, 2221–2232 (2014)
7. Mittal, K., Jain, A., Vaisla, K.S., Castillo, O., Kacprzyk, J.: A comprehensive review on type 2 fuzzy logic applications: past, present and future. Eng. Appl. Artif. Intell. **95**, 103916 (2020)
8. Ibrahim, A.A., Zhou, H.B., Tan, S.X., Zhang, C.L., Duan, J.A.: Regulated Kalman filter based training of an interval type-2 fuzzy system and its evaluation. Eng. Appl. Artif. Intell. **95**, 103867 (2020)
9. Balootaki, M.A., Rahmani, H., Moeinkhah, H., Mohammadzadeh, A.: On the Synchronization and Stabilization of fractional-order chaotic systems: Recent advances and future perspectives. Physica A **551**, 124203 (2020)
10. Sanchez, M.A., Castro, J.R., Ocegueda-Miramontes, V., Cervantes, L.: Hybrid learning for general type-2 TSK fuzzy logic systems. Algorithms **10**, 99 (2017)
11. Ontiveros, E., Melin, P., Castillo, O.: High order α-planes integration: a new approach to computational cost reduction of general Type-2 fuzzy systems. Eng. Appl. Artif. Inteligence **4**, 186–197 (2018)
12. Wu, D., Mendel, J.M.: Recommendations on designing practical interval type-2 fuzzy systems. Eng. Appl. Artif. Intell. **85**, 182–193 (2019)
13. Chiclana, F., Zhou, S.M.: Type-reduction of general type-2 fuzzy sets: the type-1 OWA approach. Int. J. Intell. Syst. **28**, 505–522 (2013)
14. Jeng, W.H.R., Yeh, C.Y., Lee, S.J.: General type-2 fuzzy neural network with hybrid learning for function approximation. In: 2009 IEEE International Conference on Fuzzy Systems, pp. 1534–1539 (2009)
15. Figueroa-García, J.C., Román-Flores, H., Chalco-Cano, Y.: Type–reduction of Interval Type–2 fuzzy numbers via the Chebyshev inequality. Fuzzy Sets Syst. (2021)
16. Castillo, O., Muhuri, P.K., Melin, P., Pulkkinen, P.: Emerging Issues and Applications of Type-2 Fuzzy Sets and Systems (2020)
17. Sahab, N., Hagras, H.: Adaptive non-singleton type-2 fuzzy logic systems: a way forward for handling numerical uncertainties in real world applications. Int. J. Comput. Commun. Control **6**, 503–529 (2011)
18. Tavana, M.R., Khooban, M.H., Niknam, T.: Adaptive PI controller to voltage regulation in power systems: STATCOM as a case study. ISA Trans. **66**, 325–334 (2017)
19. Mohammadzadeh, A., Sabzalian, M.H., Ahmadian, A., Nabipour, N.: A dynamic general type-2 fuzzy system with optimized secondary membership for online frequency regulation. ISA Trans. **112**, 150–160 (2021)
20. Torshizi, A.D., Zarandi, M.H.F.: A new cluster validity measure based on general type-2 fuzzy sets: application in gene expression data clustering. Knowl. Based Syst. **64**, 81–93 (2014)
21. Mohammadzadeh, A., Kumbasar, T.: A new fractional-order general type-2 fuzzy predictive control system and its application for glucose level regulation. Appl. Soft Comput. **91**, 106241 (2020)

22. Khooban, M.H., Vafamand, N., Liaghat, A., Dragicevic, T.: An optimal general type-2 fuzzy controller for Urban Traffic Network. ISA Trans. **66**, 335–343 (2017)
23. Zarandi, M.F., Soltanzadeh, S., Mohammadi, A., Castillo, O.: Designing a general type-2 fuzzy expert system for diagnosis of depression. Appl. Soft Comput. **80**, 329–341 (2019)
24. Carvajal, O., Melin, P., Miramontes, I., Prado-Arechiga, G.: Optimal design of a general type-2 fuzzy classifier for the pulse level and its hardware implementation. Eng. Appl. Artif. Intell. **97**, 104069 (2021)
25. Zhao, T., Liu, J., Dian, S., Guo, R., Li, S.: Sliding-mode-control-theory-based adaptive general Type-2 fuzzy neural network control for power-line inspection robots. Neurocomputing **401**, 281–294 (2020)
26. Ontiveros-Robles, E., Castillo, O., Melin, P.: Towards asymmetric uncertainty modeling in designing General Type-2 Fuzzy classifiers for medical diagnosis. Expert Syst. Appl. **183**, 115370 (2021)
27. Doctor, F., Syue, C.H., Liu, Y.X., Shieh, J.S., Iqbal, R.: Type-2 fuzzy sets applied to multivariable self-organizing fuzzy logic controllers for regulating anesthesia. Appl. Soft Comput. **38**, 872–889 (2016)
28. Geramian, A., Abraham, A.: Customer classification: a Mamdani fuzzy inference system standpoint for modifying the failure mode and effect analysis based three dimensional approach. Expert Syst. Appl., 115753 (2021)
29. Ontiveros-Robles, E., Melin, P.: A hybrid design of shadowed type-2 fuzzy inference systems applied in diagnosis problems. Eng. Appl. Artif. Intell. **86**, 43–55 (2019)
30. Ochoa, P., Castillo, O., Melin, P., Soria, J.: Differential evolution with shadowed and General Type-2 fuzzy systems for dynamic parameter adaptation in optimal design of fuzzy controllers (2021)
31. Almaraashi, M., John, R., Hopgood, A., Ahmadi, S.: Learning of interval and general type-2 fuzzy logic systems using simulated annealing: Theory and practice. Inforamtion Sci. **360**, 21–42 (2016)
32. Mendel, J.M.: General type-2 fuzzy logic systems made simple: a tutorial. IEEE Trans. Fuzzy Syst. **22**, 1162–1182 (2013)
33. http://www.vision.caltech.edu/bouguetj/calib_doc/index.html#ref
34. Carlotto, M.J.: Detecting change in images with parallax. In: Defense and Security Symposium, pp. 656719–656719. International Society for Optics and Photonics (2007)
35. Davies, E.R.: The application of machine vision to food and agriculture: a review. Imaging Sci. J. **57**(4), 197–217 (2009)
36. Demant, C., Demant, C., Streicher-Abel, B.: Industrial Image Processing. Springer (1999)
37. http://www.guioteca.com/fotografia/entendiendo-la-exposicion-en-fotografia-1%C2%AA-parte/
38. Taylor, B.N.: Guidelines for Evaluating and Expressing the Uncertainty of NIST Measurement Results. DIANE Publishing (2009)
39. Mouzouris, G.C., Mendel, J.M.: Dynamic non-singleton fuzzy logic systems for nonlinear modeling. Fuzzy Syst. IEEE Trans. **5**(2), 199–208 (1997)
40. Méndez, G.M., Dorantes, P.N.M., Mexicano, A.: Interval type-2 fuzzy logic systems optimized by central composite design to create a simplified fuzzy rule base in image processing for quality control application. Int. J. Adv. Manuf. Technol. **102**(9–12), 3757–3766 (2019)
41. Montes Dorantes, P.N., Nieto González, J.P., Praga-Alejo, R., Guajardo Cosio, K.L., Méndez, G.M.: Sistema inteligente para procesamiento de imágenes en control de calidad basado en el modelo difuso singleton tipo 1. Res. Comput. Sci. **74**, 117–130 (2014)
42. Montes Dorantes, P.N., Jiménez Gómez, M.A, Méndez, G.M., Nieto González, J.P., de la Rosa Elizondo, J.: One step models for soft computing techniques. Industrial application to image processing in quality assurance process. Int. J. Adv. Manuf. Technol. (IJAMT, Springer), **81**(5), 771–778 (2015)
43. Dorantes, P.N.M., Méndez, G.M.: Non-iterative radial basis function neural networks to quality control via image processing. IEEE Lat. Am. Trans. **13**(10), 3457–3451 (2015)

44. Dorantes, P.N.M., Mexicano, S.A., Méndez, G.M.: Modeling Type-1 singleton fuzzy logic systems using statistical parameters in foundry temperature control application. Smart Sustain. Manuf. Syst. **2**(1), 180–203 (2018)
45. Taylor, B.N., Kuyatt, C.E.: NIST, (National Institute of Standards and Technology, United States Department of Commerce Technology Administration. Technical Note 1297, Guidelines for Evaluating and Expressing the Uncertainty of NIST Measurement Results (1994)
46. Braunschweig, W.K.: ISO/BIMP, Uncertainty of Measurement (1999)

Type-2 Neutrosophic Fuzzy Bimatrix Games Based on a New Distance Measure

Shuvasree Karmakar and Mijanur Rahaman Seikh

Abstract In reality, multiple sources of information, a large volume of data, and different parameters encourage us to discuss bimatrix game problems with the type-2 fuzzy backdrop. Further, to apprehend the data in terms of three kinds of membership values: truth, indeterminacy, and falsity, in the present investigation, we aspire to depict bimatrix games with payoffs of the type-2 neutrosophic fuzzy set (T2NFS). To do this, we define the Euclidean distance measure of T2NFSs. We describe an aggregation operator of T2NFSs to aggregate information in terms of T2NFS. Then based on the proposed distance, we also define a similarity measure of T2NFSs. Later, we solve the bimatrix game utilizing the proposed similarity measure of T2NFSs. Finally, we justify the established model by offering a real-world problem to mitigate environmental pollution.

Keywords Type-2 neutrosophic fuzzy set · Algebraic operator · Normalized Euclidean distance · Similarity measure · Bimatrix games · Pollution mitigating problem

1 Introduction

Nash [15] presented the bimatrix game as a unique tool to design competitive problems in the non-cooperative game theory. Initially, the game theorists considered numeric values for the payoffs of a bi-matrix game. Later, considering the insufficiency in the information, bimatrix games experienced their extensions in fuzzy background. Recently, numerous initiatives took place in the fuzzy bimatrix games [11, 25]. Li and Li [9] formulated bimatrix games taking fuzzy entities in the credibility space. With a mean-area-based ranking of the intuitionistic fuzzy sets (IFSs), An et al. [1] performed a method with a non-linear programming problem (NLPP) for solving bimatrix games. Brikaa et al. [3] conceived a tourism planning strategy

S. Karmakar · M. R. Seikh (✉)
Department of Mathematics, Kazi Nazrul University, Asansol 713340, India
e-mail: mrseikh@ymail.com

© The Author(s), under exclusive license to Springer Nature Switzerland AG 2023   217
O. Castillo and A. Kumar (eds.), *Recent Trends on Type-2 Fuzzy Logic Systems: Theory, Methodology and Applications*, Studies in Fuzziness and Soft Computing 425,
https://doi.org/10.1007/978-3-031-26332-3_14

model in the light of the rough interval bimatrix game. Li and Tu [12] executed a solution methodology for solving IF bimatrix games. Using the concept of the proportion mix of the possibility and necessity expectation, Khan and Mehra [8] solved bimatrix games with IF payoffs. To solve bimatrix games with trapezoidal IF numbers Li and Yang [10] used a difference-based ranking approach. Yang et al. [24] used NLPP approach to resolve bimatrix games with IF backdrops. Furthermore, IFSs emerged in the bimatrix games by numerous researchers [14, 20, 21] in several ways. Karmakar and Seikh [7] developed dense NLPP for solving bimatrix games having dense fuzzy payoffs. Bhaumik et al. [2] developed a neutrosophic bimatrix game depend on (α, β, γ)−cut sets.

Reality allows decision-makers to depict their decisions in the shape of the type-2 fuzzy sets (T2FSs). T2FS is efficient in sketching out the exact membership value of the intake information due to the deficiency of intentness, time, etc. Furthermore, occasionally, due to impreciseness, inconsistency, and incomplete information, the membership value of the intake data experiences variation. In that case, we have three kinds of membership values: truth, indeterminacy, and falsity. Moreover, in some cases, due to the large volume of data and multiple sources of information, a two-dimensional fuzzy set is not enough to portray the situation. In 2020, Karaasalan and Hunu [5] introduced the notion of type-2 neutrosophic fuzzy sets (T2NFSs) with few of its properties to cover this gap. Karaasalan and Hunu [5] applied it to solve a multi-attribute decision making (MADM) problem. Ozlu and Karaslan [17] executed a hybrid similarity measure of T2NFSs and used it to solve multi-criteria decision making (MCDM) problems.

In the recent past, few works have been developed in the theory of matrix games with type-2 fuzzy backdrops. Roy and Bhaumik [18] analyzed the water management problem in the light of type-2 intuitionistic fuzzy matrix games. Karmakar et al. [6] defined a Minkowski distance and similarity measures of type-2 intuitionistic fuzzy sets based on the Hausdorff metric and justified its applicability to the matrix game. Seikh et al. [22] developed a new defuzzification method for type-2 fuzzy variables and utilizing this method they solved the matrix game problem with payoffs of type-2 fuzzy variables. Roy and Maiti [19] proposed a different kind of reduction method of type-2 fuzzy variable and used it to solve Stackelberg games.

In reality, due to the lack of complete precise information, extensive sources of information, etc., decision-makers (or the players for a game) fail to assess their payoffs in terms of type-1 fuzzy sets. In this chapter, we portray a bimatrix game with T2NF background to cope with such a situation. To the best of our knowledge, this is the first attempt to discuss a bimatrix game in the T2NF environment. We execute a quadratic type-2 neutrosophic fuzzy programming problem to solve the game. But, it's a tedious job to handle such a programming problem with T2NFSs. We define algebraic operators of T2NFSs to aggregate the intake information. We also represent the normalized Euclidean distance of T2FNs, and in that context, we define a similarity index of T2NFSs. Rather than the traditional way of solving, we extend the notion of the composite relative degree of similarity to the positive ideal solution in the T2NF scenario and solve the bimatrix game. We illustrate a problem to alleviate environmental pollution to justify the applicability of the presented model.

The contributions of the present chapter are as follows.

- The present chapter explores the solution methodology for a bimatrix game with T2NF payoffs considering the truth, indeterminacy, and falsity of the imprecise information. This is the first attempt in the bimatrix game theory considering type-2 fuzzy neutrosophic payoffs.
- Despite the traditional method, we develop the solution using the concept of players' best choice towards the information. This method helps us to avoid the loss of information which generally occurs during type reduction.

We design the rest of the chapter in the following manner. We recall basic definitions associated with T2FNSs in Sect. 2. Section 3 defines algebraic operations on T2NFSs to aggregate the information. Section 4 proposed the normalized Euclidean distance measure and a similarity index of T2NFSs, which helps make a ranking order of T2NF information. Section 5 establishes a bimatrix game model in the T2NF scenario and its solution approach. To justify the applicability of the proposed methodology, Sect. 6 presents a real-world application. Finally, Sect. 7 concludes the discussion.

2 Preliminaries

Here, we recall some definitions to establish the subsequent discussion.

Definition 1 **NFS** ([23]) Let $\mathcal{N}$ be a non-empty set. A set $\tilde{A} = \{\langle x, \tau_A(x), \iota_A(x), \xi_A(x) \rangle : x \in \mathcal{N}\}$ is called a neutrosophic fuzzy set or NFS where $\tau_A(x), \iota_A(x)$ and $\xi_A(x)$ takes values from the interval $[0, 1]$ and $0 \leq \tau_A(x) + \iota_A(x) + \xi_A(x)$. $\tau_A(x), \iota_A(x)$ and $\xi_A(x)$ are termed as the truth, indeterminacy and falsity membership values, respectively.

Definition 2 **T2FS** ([13]): Suppose U be the universal set. A T2FS $\tilde{\tilde{A}}$ is defined in terms of its primary membership and secondary membership functions κ and ζ as $\tilde{\tilde{A}} = \{(x, \kappa, \zeta_A) : x \in U, \kappa(x)) \in j_x \subseteq [0, 1]\}$.

Definition 3 **T2NFS** ([5]): Let $\mathcal{T}$ be a non-empty set. Then we can define a T2NFS $\tilde{\tilde{A}}$ in terms of its truth membership function $\tau_A(x)$, and indeterminacy membership function $\iota_A(x)$ and falsity membership function $\xi_A(x)$ and is denoted as $\tilde{\tilde{A}} = \{\langle x, \tau_A(x), \iota_A(x), \xi_A(x) \rangle : x \in \mathcal{T}\}$ where $\tau_A(x) = (\tau_{pA}, \tau_{sA}), \iota_A(x) = (\iota_{pA}, \iota_{sA})$ and $\xi_A(x) = (\xi_{pA}, \xi_{sA})$ are type-2 fuzzy elements. Also, $0 \leq \tau_{pA} + \iota_{pA} + \xi_{pA} \leq 1$ and $0 \leq \tau_{sA} + \iota_{sA} + \xi_{sA} \leq 1$.

For convenience, we represent a T2NFS as $\tilde{\tilde{A}} = \langle (\tau_{pA}, \tau_{sA}), (\iota_{pA}, \iota_{sA}), (\xi_{pA}, \xi_{sA}) \rangle$ throughout the discussion.

Definition 4 **Null T2NFS:** A T2NFS $\tilde{\tilde{\Phi}}$ is called a null T2NFS if all of its truth membership function $\tau_{\Phi}(x)$, and indeterminacy membership function $\iota_{\Phi}(x)$ and falsity membership function $\xi_{\Phi}(x)$ take the value zero. Thus, we can represent a null T2NFS as $\tilde{\tilde{\Phi}} = \langle (0, 0), (0, 0), (0, 0) \rangle$.

3   Aggregation Operator of T2NFS

In several situation, we have to aggregate the information. This section develops the algebraic operator of T2NFSs.

Let us consider three T2NFSs, $\tilde{\tilde{A}}, \tilde{\tilde{B}}, \tilde{\tilde{C}} \in T(U)$ having form $\tilde{\tilde{A}} = \langle(\tau_{pA}, \tau_{sA});$
$(\iota_{pA}, \iota_{sA}); (\xi_{pA}, \xi_{sA})\rangle$, $\tilde{\tilde{B}} = \langle(\tau_{pB}, \tau_{sB}); (\iota_{pB}, \iota_{sB}); (\xi_{pB}, \xi_{sB})\rangle$ and $\tilde{\tilde{C}} = \langle(\tau_{pC}, \tau_{sC});$
$(\iota_{pC}, \iota_{sC}); (\xi_{pC}, \xi_{sC})\rangle$. Then the algebraic operations on the T2NFSs are defined as follows.

1. $\tilde{\tilde{A}} \oplus_1 \tilde{\tilde{B}} = \left\{\left\langle\left(\tau_{pA} + \tau_{pB} - \tau_{pA}\tau_{pB}, \tau_{sA} + \tau_{sB} - \tau_{sA}\tau_{sB}\right); \left(\iota_{pA}\iota_{pB}, \iota_{sA}\iota_{sB}\right);\right.\right.$
$\left.\left.\left(\xi_{pA}\xi_{pB}, \xi_{sA}\xi_{sB}\right)\right\rangle\right\}$.

2. $\tilde{\tilde{A}} \otimes_1 \tilde{\tilde{B}} = \left\{\left\langle\left(\tau_{pA}\tau_{pB}, \tau_{sA}\tau_{sB}\right); \left(\iota_{pA} + \iota_{pB} - \iota_{pA}\iota_{pB}, \iota_{sA} + \iota_{sB} - \iota_{sA}\iota_{sB}\right);\right.\right.$
$\left.\left.\left(\xi_{pA} + \xi_{pB} - \xi_{pA}\xi_{pB}, \xi_{sA} + \xi_{sB} - \xi_{sA}\xi_{sB}\right)\right\rangle\right\}$.

3. $h\tilde{\tilde{A}} = \left\{\left\langle\left(\dfrac{1 - (1 - \tau_{pA})^h}{1 + (1 - \tau_{pA})^h}, \dfrac{1 - (1 - \tau_{sA})^h}{1 + (1 - \tau_{sA})^h}\right); \left((\iota_{pA})^h, (\iota_{sA})^h\right);\right.\right.$
$\left.\left.\left((\xi_{pA})^h, (\xi_{sA})^h\right)\right\rangle\right\}$.

4. $\tilde{\tilde{A}}^h = \left\{\left\langle\left((\tau_{pA})^h, (\tau_{sA})^h\right); \left(\dfrac{1 - (1 - \iota_{pA})^h}{1 + (1 - \iota_{pA})^h}, \dfrac{1 - (1 - \iota_{sA})^h}{1 + (1 - \iota_{sA})^h}\right); \left(\dfrac{1 - (1 - \xi_{pA})^h}{1 + (1 - \xi_{pA})^h},\right.\right.\right.\right.$
$\left.\left.\left.\left.\dfrac{1 - (1 - \xi_{sA})^h}{1 + (1 - \xi_{sA})^h}\right)\right)\right\rangle\right\}$.

The operators described above of T2NFSs assist us in establishing the following theorem.

Theorem 1 *Let us assume a collection of n numbers of T2NFSs $\tilde{\tilde{A}}_t$ ($t = 1, 2, ..., n,$).*
Also, suppose $\omega_t's$ indicates the weight associated to each T2NFS $\tilde{\tilde{A}}_t$, with $\sum_{t=1}^{n} \omega_t = 1$.
Thus, we can calculate the aggregated weighted algebraic sum (AWAS) as follows.

$$\oplus_{t=1}^{n} \omega_t \tilde{\tilde{A}}_t = \left\langle\left(\frac{1 - \prod_{t=1}^{n}(1 - \tau_{pA_t})^{\omega_t}}{1 + \prod_{t=1}^{n}(1 - \tau_{pA_t})^{\omega_t}}, \frac{1 - \prod_{t=1}^{n}(1 - \tau_{sA_t})^{\omega_t}}{1 + \prod_{t=1}^{n}(1 - \tau_{sA_t})^{\omega_t}}\right);\right.$$
$$\left.\left(\prod_{t=1}^{n}(\iota_{pA_t})^{\omega_t}; \prod_{t=1}^{n}(\iota_{sA_t})^{\omega_t}\right); \left(\prod_{t=1}^{n}(\xi_{pA_t})^{\omega_t}; \prod_{t=1}^{n}(\xi_{sA_t})^{\omega_t}\right)\right\rangle. \quad (1)$$

Proof Suppose $\tilde{\tilde{A}}_1$ and $\tilde{\tilde{A}}_2$ be two T2NFSs. ω_1 and ω_2 are two weight vectors associated to $\tilde{\tilde{A}}_1$ and $\tilde{\tilde{A}}_2$, respectively. Then by algebraic sum and scalar product, we have

$$\omega_1 \tilde{\tilde{A}}_1 \oplus \omega_2 \tilde{\tilde{A}}_2 = \omega_1 \langle (\tau_{pA_1}, \tau_{sA_1}); (\iota_{pA_1}, \iota_{sA_1}); (\xi_{pA_1}, \xi_{sA_1}) \rangle \oplus$$
$$\omega_2 \langle (\tau_{pA_2}, \tau_{sA_2}); (\iota_{pA_2}, \iota_{sA_2}); (\xi_{pA_2}, \xi_{sA_2}) \rangle$$

$$= \left\langle \left(\frac{1 - (1 - \tau_{pA_1})^{\omega_1}}{1 + (1 - \tau_{pA_1})^{\omega_1}}, \frac{1 - (1 - \tau_{sA_1})^{\omega_1}}{1 + (1 - \tau_{sA_1})^{\omega_1}} \right);$$
$$\left((\iota_{pA_1})^{\omega_1}; (\iota_{sA_1})^{\omega_1} \right); \left((\xi_{pA_1})^{\omega_1}; (\xi_{sA_1})^{\omega_1} \right) \right\rangle \oplus$$
$$\left\langle \left(\frac{1 - (1 - \tau_{pA_2})^{\omega_2}}{1 + (1 - \tau_{pA_2})^{\omega_2}}, \frac{1 - (1 - \tau_{sA_2})^{\omega_2}}{1 + (1 - \tau_{sA_2})^{\omega_2}} \right);$$
$$\left((\iota_{pA_2})^{\omega_2}; (\iota_{sA_2})^{\omega_2} \right); \left((\xi_{pA_2})^{\omega_2}; (\xi_{sA_2})^{\omega_2} \right) \right\rangle$$

$$= \left\langle \left(\frac{1 - (1 - \tau_{pA_1})^{\omega_1} (1 - \tau_{pA_2})^{\omega_2}}{1 + (1 - \tau_{pA_1})^{\omega_1} (1 - \tau_{pA_2})^{\omega_2}}, \frac{1 - (1 - \tau_{sA_1})^{\omega_1} (1 - \tau_{sA_2})^{\omega_2}}{1 + (1 - \tau_{sA_1})^{\omega_1} (1 - \tau_{sA_2})^{\omega_2}} \right);$$
$$\left((\iota_{pA_1})^{\omega_1} (\iota_{pA_2})^{\omega_2}; (\iota_{sA_1})^{\omega_1} (\iota_{sA_2})^{\omega_2} \right);$$
$$\left((\xi_{pA_1})^{\omega_1} (\xi_{pA_2})^{\omega_2}; (\xi_{sA_1})^{\omega_1} (\xi_{sA_2})^{\omega_2} \right) \right\rangle.$$

Consequently, the mathematical statement is valid for $t = 2$.
Let us assume that the statement is valid for $t = k$(some natural number), i.e.,

$$\oplus_{t=1}^{k} \omega_t \tilde{\tilde{A}}_t = \left\langle \left(\frac{1 - \prod_{t=1}^{k} (1 - \tau_{pA_t})^{\omega_t}}{1 + \prod_{t=1}^{k} (1 - \tau_{pA_t})^{\omega_t}}, \frac{1 - \prod_{t=1}^{k} (1 - \tau_{sA_t})^{\omega_t}}{1 + \prod_{t=1}^{k} (1 - \tau_{sA_t})^{\omega_t}} \right); \right.$$
$$\left. \left(\prod_{t=1}^{k} (\iota_{pA_t})^{\omega_t}; \prod_{t=1}^{k} (\iota_{sA_t})^{\omega_t} \right); \left(\prod_{t=1}^{k} (\xi_{pA_t})^{\omega_t}; \prod_{t=1}^{k} (\xi_{sA_t})^{\omega_t} \right) \right\rangle.$$

Now for $t = k + 1$, we have

$$\oplus_{t=1}^{k+1} \omega_t \tilde{\tilde{A}}_t = \left(\oplus_{t=1}^{k} \omega_t \tilde{\tilde{A}}_t \right) \oplus \omega_{k+1} \tilde{\tilde{A}}_{k+1}$$

$$= \left\langle \left(\frac{1 - \prod_{t=1}^{k} (1 - \tau_{pA_t})^{\omega_t}}{1 + \prod_{t=1}^{k} (1 - \tau_{pA_t})^{\omega_t}}, \frac{1 - \prod_{t=1}^{k} (1 - \tau_{sA_t})^{\omega_t}}{1 + \prod_{t=1}^{k} (1 - \tau_{sA_t})^{\omega_t}} \right); \right.$$
$$\left(\prod_{t=1}^{k} (\iota_{pA_t})^{\omega_t}; \prod_{t=1}^{k} (\iota_{sA_t})^{\omega_t} \right); \left(\prod_{t=1}^{k} (\xi_{pA_t})^{\omega_t}; \prod_{t=1}^{k} (\xi_{sA_t})^{\omega_t} \right) \right\rangle \oplus$$
$$\left\langle \left(\frac{1 - (1 - \tau_{pA_{k+1}})^{\omega_{k+1}}}{1 + (1 - \tau_{pA_{k+1}})^{\omega_{k+1}}}, \frac{1 - (1 - \tau_{sA_{k+1}})^{\omega_{k+1}}}{1 + (1 - \tau_{sA_{k+1}})^{\omega_{k+1}}} \right); \right.$$
$$\left. \left((\iota_{pA_{k+1}})^{\omega_{k+1}}; (\iota_{sA_{k+1}})^{\omega_{k+1}} \right); \left((\xi_{pA_{k+1}})^{\omega_{k+1}}; (\xi_{sA_{k+1}})^{\omega_{k+1}} \right) \right\rangle$$

$$= \left\langle \left(\frac{1 - \prod_{t=1}^{k+1} (1 - \tau_{pA_t})^{\omega_t}}{1 + \prod_{t=1}^{k+1} (1 - \tau_{pA_t})^{\omega_t}}, \frac{1 - \prod_{t=1}^{k+1} (1 - \tau_{sA_t})^{\omega_t}}{1 + \prod_{t=1}^{k+1} (1 - \tau_{sA_t})^{\omega_t}} \right); \right.$$
$$\left. \left(\prod_{t=1}^{k+1} (\iota_{pA_t})^{\omega_t}; \prod_{t=1}^{k+1} (\iota_{sA_t})^{\omega_t} \right); \left(\prod_{t=1}^{k+1} (\xi_{pA_t})^{\omega_t}; \prod_{t=1}^{k+1} (\xi_{sA_t})^{\omega_t} \right) \right\rangle.$$

Accordingly, the principle of mathematical induction concludes the theorem. □

We can easily verify that the AWAS of T2NFSs meets the following properties:

(i) **Monotonicity**: For two families of T2NFSs $\tilde{\tilde{B}}_t$ and $\tilde{\tilde{A}}_t$, if $\tilde{\tilde{B}}_t \leq \tilde{\tilde{A}}_t$, for $t = 1, 2, ..., n$, then $\oplus_{t=1}^n \omega_t \tilde{\tilde{B}}_t \leq \oplus_{t=1}^n \omega_t \tilde{\tilde{A}}_t$.

(ii) **Idempotency**: We have $\oplus_{t=1}^n \omega_t \tilde{\tilde{A}}_t = \tilde{\tilde{A}}$, if $\tilde{\tilde{A}}_t = \tilde{\tilde{A}}$ for $t = 1, 2, ...n$.

(iii) **Boundedness**: $\min_{1 \leq t \leq n} \tilde{\tilde{A}}_t \leq \oplus_{t=1}^n \omega_t \tilde{\tilde{A}}_t \leq \max_{1 \leq t \leq n} \tilde{\tilde{A}}_t$.

4 Novel Distance Measure and Similarity Index of T2NFSs

This section introduces the normalized Euclidean distance of T2NFSs depending on the Hausdorff metric. Moreover, we present a similarity index of T2NFSs, which help us to make an order of T2NFSs to solve various problems in reality.

4.1 Normalized Euclidean Distance

Suppose $\tilde{\tilde{A}} = \langle (\tau_{pA(r)}, \tau_{sA(r)}); (\iota_{pA(r)}, \iota_{sA(r)}); (\xi_{pA(r)}, \xi_{sA(r)}) \rangle$ and $\tilde{\tilde{B}} = \langle (\tau_{pB(r)}, \tau_{sB(r)}); (\iota_{pB(r)}, \iota_{sB(r)}); (\xi_{pB(r)}, \xi_{sB(r)}) \rangle$ be two T2NFSs with $r = 1, 2, \ldots, n$. Then we can determine the normalized Euclidean distance of $\tilde{\tilde{A}}$ and $\tilde{\tilde{B}}$ depending on Hausdorff metric as

$$d_e(\tilde{\tilde{A}}, \tilde{\tilde{B}}) = \frac{1}{3n} \sum_{r=1}^n \left\{ \left(|\tau_{pA(r)} - \tau_{pB(r)}|^2 + |\iota_{pA(r)} - \iota_{pB(r)}|^2 + |\xi_{pA(r)} - \xi_{pB(r)}|^2 \right) \vee \left(|\tau_{sA(r)} - \tau_{sB(r)}|^2 + |\iota_{sA(r)} - \iota_{sB(r)}|^2 + |\xi_{sA(r)} - \xi_{sB(r)}|^2 \right) \right\}. \tag{2}$$

Theorem 2 *Suppose $\mathcal{T}$ be the family of all T2NFSs. The normalized Euclidean distance of two T2NFSs $\tilde{\tilde{A}}, \tilde{\tilde{B}} \in \mathcal{T}$, depicted in Eq. (2) accomplishes the below-mentioned properties.*

$P_d(1)$: $d_e(\tilde{\tilde{A}}, \tilde{\tilde{B}}) \in [0, 1]$, $\forall \tilde{\tilde{A}}, \tilde{\tilde{B}} \in \mathcal{T}$.
$P_d(2)$: $d_e(\tilde{\tilde{A}}, \tilde{\tilde{B}}) = 0 \Leftrightarrow \tilde{\tilde{A}} = \tilde{\tilde{B}}$ for $\tilde{\tilde{A}}, \tilde{\tilde{B}} \in \mathcal{T}$.
$P_d(3)$: $d_e(\tilde{\tilde{A}}, \tilde{\tilde{B}}) = d_e(\tilde{\tilde{B}}, \tilde{\tilde{A}})$, $\forall \tilde{\tilde{A}}, \tilde{\tilde{B}} \in \mathcal{T}$.
$P_d(4)$: If $d_e(\tilde{\tilde{A}}, \tilde{\tilde{B}}) = d_e(\tilde{\tilde{A}}, \tilde{\tilde{C}}) = 0$ then $\forall \tilde{\tilde{A}}, \tilde{\tilde{B}}, \tilde{\tilde{C}} \in \mathcal{T}$, $d_e(\tilde{\tilde{B}}, \tilde{\tilde{C}}) = 0$.

Proof

$P_d(1)$: We already have $\tau_p, \tau_s, \iota_p, \iota_s, \xi_p, \xi_s \in [0, 1]$. So, $|\tau_{pA(r)} - \tau_{pB(r)}|^2$, $|\tau_{sA(r)} - \tau_{sB(r)}|^2$, $|\iota_{pA(r)} - \iota_{pB(r)}|^2$, $|\iota_{sA(r)} - \iota_{sB(r)}|^2$, $|\xi_{pA(r)} - \xi_{pB(r)}|^2$,

$|\xi_{sA(r)} - \xi_{sB(r)}|^2$ must be some positive quantities. Thus we have, $d_e(\tilde{\tilde{A}}, \tilde{\tilde{B}}) \geq 0$. Also, $|\tau_{pA(r)} - \tau_{pB(r)}|^2, |\iota_{pA(r)} - \iota_{pB(r)}|^2, |\xi_{pA(r)} - \xi_{pB(r)}|^2 \leq 1$, which shows $|\tau_{pA(r)} - \tau_{pB(r)}|^2 + |\iota_{pA(r)} - \iota_{pB(r)}|^2 + |\xi_{pA(r)} - \xi_{pB(r)}|^2 \leq 3$ or we may write $\frac{1}{3}\left(|\tau_{pA(r)} - \tau_{pB(r)}|^2 + |\iota_{pA(r)} - \iota_{pB(r)}|^2 + |\xi_{pA(r)} - \xi_{pB(r)}|^2\right) \leq 1$.

Similarly, we have $\frac{1}{3}\left(|\tau_{sA(r)} - \tau_{sB(r)}|^2 + |\iota_{sA(r)} - \iota_{sB(r)}|^2 + |\xi_{sA(r)} - \xi_{sB(r)}|^2\right) \leq 1$. Thus, we write $d_e(\tilde{\tilde{A}}, \tilde{\tilde{B}}) \leq 1$. So, we can conclude $d_e(\tilde{\tilde{A}}, \tilde{\tilde{B}}) \in [0, 1]$.

$\mathbf{P}_d(2)$: Let us suppose that $d_e(\tilde{\tilde{A}}, \tilde{\tilde{B}}) = 0$. This indicates $\frac{1}{3n} \sum_{r=1}^{n} \left\{\left(|\tau_{pA(r)} - \tau_{pB(r)}|^2 + |\iota_{pA(r)} - \iota_{pB(r)}|^2 + |\xi_{pA(r)} - \xi_{pB(r)}|^2\right) \vee \left(|\tau_{sA(r)} - \tau_{sB(r)}|^2 + |\iota_{sA(r)} - \iota_{sB(r)}|^2 + |\xi_{sA(r)} - \xi_{sB(r)}|^2\right)\right\} = 0$. Such relation is possible only if both the quantities $|\tau_{pA(r)} - \tau_{pB(r)}|^2 + |\iota_{pA(r)} - \iota_{pB(r)}|^2 + |\xi_{pA(r)} - \xi_{pB(r)}|^2$ and $|\tau_{sA(r)} - \tau_{sB(r)}|^2 + |\iota_{sA(r)} - \iota_{sB(r)}|^2 + |\xi_{sA(r)} - \xi_{sB(r)}|^2$ are equal to 0. This is occurred if we have $\tau_{pA(r)} = \tau_{pB(r)}$, $\iota_{pA(r)} = \iota_{pB(r)}$, $\xi_{pA(r)} = \xi_{pB(r)}|^2$, $\tau_{sA(r)} = \tau_{sB(r)}$, $\iota_{sA(r)} = \iota_{sB(r)}$, $\xi_{sA(r)} = \xi_{sB(r)}$, i.e., $\tilde{\tilde{A}} = \tilde{\tilde{B}}$. Thus we may conclude $d_e(\tilde{\tilde{A}}, \tilde{\tilde{B}}) = 0$ or vice-versa.

$\mathbf{P}_d(3)$: From Eq. (2), we have

$$d_e(\tilde{\tilde{A}}, \tilde{\tilde{B}}) = \frac{1}{3n} \sum_{r=1}^{n} \left\{\left(|\tau_{pA(r)} - \tau_{pB(r)}|^2 + |\iota_{pA(r)} - \iota_{pB(r)}|^2 + |\xi_{pA(r)} - \xi_{pB(r)}|^2\right) \vee \right.$$
$$\left(|\tau_{sA(r)} - \tau_{sB(r)}|^2 + |\iota_{sA(r)} - \iota_{sB(r)}|^2 + |\xi_{sA(r)} - \xi_{sB(r)}|^2\right)\right\}$$
$$= \frac{1}{3n} \sum_{r=1}^{n} \left\{\left(|\tau_{pB(r)} - \tau_{pA(r)}|^2 + |\iota_{pB(r)} - \iota_{pA(r)}|^2 + |\xi_{pB(r)} - \xi_{pA(r)}|^2\right) \vee \right.$$
$$\left(|\tau_{sB(r)} - \tau_{sA(r)}|^2 + |\iota_{sB(r)} - \iota_{sA(r)}|^2 + |\xi_{sB(r)} - \xi_{sA(r)}|^2\right)\right\}$$
$$= d_e(\tilde{\tilde{B}}, \tilde{\tilde{A}}).$$

$\mathbf{P}_d(4)$: Property $\mathbf{P}_d(3)$: immediately proves the present property. $\qquad\square$

4.2 Similarity Index of T2NFSs

The similarity index of a fuzzy set has broad application for dealing with numerous problems in reality. Here, we describe a similarity index of T2NFSs. This similarity index is built up based on the normalized Euclidean distance of T2NFSs.

Suppose $\tilde{\tilde{A}}, \tilde{\tilde{B}} \in \mathcal{T}$ be two T2NFSs. Then we can define the similarity index of $\tilde{\tilde{A}}$ and $\tilde{\tilde{B}}$ as

$$S_e(\tilde{\tilde{A}}, \tilde{\tilde{B}}) = \frac{1 - d_e(\tilde{\tilde{A}}, \tilde{\tilde{B}})}{1 + d_e(\tilde{\tilde{A}}, \tilde{\tilde{B}})}. \tag{3}$$

We can easily verify that the similarity index of two T2NFSs $\tilde{\tilde{A}}$, and $\tilde{\tilde{B}}$, $S_e(\tilde{\tilde{A}}, \tilde{\tilde{B}})$ ensures the below-mentioned axioms.

P$_s$(1): $S_e(\tilde{\tilde{A}}, \tilde{\tilde{B}}) \in [0, 1]$, for any two T2NFSs $\tilde{\tilde{A}}$ and $\tilde{\tilde{B}}$.

P$_s$(2): $S_e(\tilde{\tilde{A}}, \tilde{\tilde{B}}) = 1 \Leftrightarrow \tilde{\tilde{A}} = \tilde{\tilde{B}}$ for two T2NFSs $\tilde{\tilde{A}}$ and $\tilde{\tilde{B}}$.

P$_s$(3): $S_e(\tilde{\tilde{A}}, \tilde{\tilde{B}}) = S_e(\tilde{\tilde{B}}, \tilde{\tilde{A}})$, for any two T2NFSs $\tilde{\tilde{A}}$ and $\tilde{\tilde{B}}$.

P$_d$(4): If $\tilde{\tilde{A}} \subseteq \tilde{\tilde{B}} \subseteq \tilde{\tilde{C}}$, then $S_e(\tilde{\tilde{A}}, \tilde{\tilde{B}}) \geq S_e(\tilde{\tilde{A}}, \tilde{\tilde{C}})$ and $S_e(\tilde{\tilde{B}}, \tilde{\tilde{C}}) \geq S_e(\tilde{\tilde{A}}, \tilde{\tilde{C}})$ for any three T2NFSs $\tilde{\tilde{A}}$, $\tilde{\tilde{B}}$ and $\tilde{\tilde{C}} \in \mathcal{T}$.

Now, based on the proposed similarity measure of T2NFSs, we incarnate the following subsection.

4.3 Ranking Order of T2NFSs

To set the order of T2NFSs, we employ the proposed similarity index of the supplied T2NFSs to the null T2NFS $\tilde{\tilde{\Phi}}$. Suppose $\tilde{\tilde{A}}$, and $\tilde{\tilde{B}}$ be two T2NFSs. Also, $S_e(\tilde{\tilde{A}}, \tilde{\tilde{\Phi}})$ and $S_e(\tilde{\tilde{B}}, \tilde{\tilde{\Phi}})$ are the similarity index of the T2NFSs $\tilde{\tilde{A}}$ and $\tilde{\tilde{\Phi}}$, respectively. Now, we can make the ranking order of $\tilde{\tilde{A}}$ and $\tilde{\tilde{\Phi}}$ using the undermentioned rules:

(i) $\tilde{\tilde{A}} > \tilde{\tilde{\Phi}}$ if $S_e(\tilde{\tilde{A}}, \tilde{\tilde{\Phi}}) < S_e(\tilde{\tilde{B}}, \tilde{\tilde{\Phi}})$

(ii) $\tilde{\tilde{A}} < \tilde{\tilde{\Phi}}$ if $S_e(\tilde{\tilde{A}}, \tilde{\tilde{\Phi}}) > S_e(\tilde{\tilde{B}}, \tilde{\tilde{\Phi}})$

(iii) $\tilde{\tilde{A}} = \tilde{\tilde{\Phi}}$ if $S_e(\tilde{\tilde{A}}, \tilde{\tilde{\Phi}}) = S_e(\tilde{\tilde{B}}, \tilde{\tilde{\Phi}})$.

Example 1 Let us consider four T2NFSs.

$$\tilde{\tilde{A}} = \Big\{\big\langle(0.1, 0.4); (0.1, 0.3); (0.2, 0.3)\big\rangle, \big\langle(0.3, 0.4); (0.2, 0.1); (0.1, 0.1)\big\rangle,$$
$$\big\langle(0.2, 0.1); (0.3, 0.3); (0.5, 0.2)\big\rangle\Big\}$$

$$\tilde{\tilde{B}} = \Big\{\big\langle(0.5, 0.2); (0.1, 0.4); (0.1, 0.1)\big\rangle, \langle(0.5, 0.6); (0.3, 0.1); (0.4, 0.1)\rangle,$$
$$\big\langle(0.2, 0.1); (0.1, 0.1); (0.6, 0.2)\big\rangle\Big\}$$

$$\tilde{\tilde{C}} = \Big\{\big\langle(0.2, 0.3); (0.4, 0.4); (0.1, 0.2)\big\rangle, \big\langle(0.3, 0.3); (0.2, 0.4); (0.1, 0.1)\big\rangle,$$
$$\big\langle(0.3, 0.4); (0.1, 0.3); (0.2, 0.2)\big\rangle\Big\}$$

$$\tilde{\tilde{D}} = \Big\{\big\langle(0.5, 0.2); (0.1, 0.1); (0.2, 0.4)\big\rangle, \big\langle(0.5, 0.6); (0.2, 0.2); (0.1, 0.3)\big\rangle,$$
$$\big\langle(0.3, 0.5); (0.1, 0.1); (0.2, 0.1)\big\rangle\Big\}$$

Now, utilizing Eq. (3), we calculate the similarity index of the aforementioned T2NFSs as, $S_e(\tilde{\tilde{A}}, \tilde{\tilde{\Phi}}) = 0.82$, $S_e(\tilde{\tilde{B}}, \tilde{\tilde{\Phi}}) = 0.77$, $S_e(\tilde{\tilde{C}}, \tilde{\tilde{\Phi}}) = 0.83$ and $S_e(\tilde{\tilde{D}}, \tilde{\tilde{\Phi}}) = 0.79$. Thus we assign the ranking order of $\tilde{\tilde{A}}$, $\tilde{\tilde{B}}$, $\tilde{\tilde{C}}$ and $\tilde{\tilde{D}}$ as $\tilde{\tilde{B}} > \tilde{\tilde{D}} > \tilde{\tilde{A}} > \tilde{\tilde{C}}$.

5 Bimatrix Game with Type-2 Neutrosophic Payoffs

Consider a bimatrix game with T2NF payoffs. Suppose, $\mathcal{P}$ and $\mathcal{Q}$ be two players involved in this bimatrix game. Suppose $\Theta_m = (a_1, a_2, \ldots, a_m)$ be the set of pure strategies for Player $\mathcal{P}$ and $\Omega_n = (b_1, b_2, \ldots, b_n)$ be the same for Player $\mathcal{Q}$. Moreover, Y_m and Z_n are the sets of mixed strategies for Players $\mathcal{P}$ and $\mathcal{Q}$, respectively, where $Y_m = \{\mathbf{y} = (y_1, y_2, \ldots, y_m) : \sum_{j=1}^{m} y_j = 1, y_j \geq 0\}$ and $Z_n = \{\mathbf{z} = (z_1, z_2, \ldots, z_n) : \sum_{k=1}^{n} z_k = 1, z_k \geq 0\}$. Suppose Player $\mathcal{P}$ chooses $a_j (\in \Theta_m)$ and $\mathcal{Q}$ chooses $b_k (\in \Omega_n)$ as their pure strategies to maximize their profit. Then at the situation $\langle a_j, b_k \rangle$, we can estimate gains for Players $\mathcal{P}$ and $\mathcal{Q}$ in the form of T2NFSs as $\tilde{\tilde{\Gamma}}_{jk}$ and $\tilde{\tilde{\Upsilon}}_{jk}$, respectively, where $\tilde{\tilde{\Gamma}}_{jk} = \left\langle \left(\tau_{p\Gamma}^{jk}, \tau_{s\Gamma}^{jk} \right); \left(\iota_{p\Gamma}^{jk}, \iota_{s\Gamma}^{jk} \right); \left(\xi_{p\Gamma}^{jk}, \xi_{s\Gamma}^{jk} \right) \right\rangle$ and $\tilde{\tilde{\Upsilon}}_{jk} = \left\langle \left(\tau_{p\Upsilon}^{jk}, \tau_{s\Upsilon}^{jk} \right); \left(\iota_{p\Upsilon}^{jk}, \iota_{s\Upsilon}^{jk} \right); \left(\xi_{p\Upsilon}^{jk}, \xi_{s\Upsilon}^{jk} \right) \right\rangle$. Thus we may express the payoff matrices for Players $\mathcal{P}$ and $\mathcal{Q}$, respectively, as $\tilde{\tilde{\Gamma}} = (\tilde{\tilde{\Gamma}}_{jk})_{m \times n}$ and $\tilde{\tilde{\Upsilon}} = (\tilde{\tilde{\Upsilon}}_{jk})_{m \times n}$. Consequently, we may write the bimatrix game as $\tilde{\tilde{B}} = \left(Y_m, Z_n; \tilde{\tilde{\Gamma}}, \tilde{\tilde{\Upsilon}} \right)$.

If Player $\mathcal{P}$ opts $\mathbf{y} \in Y_m$ and Player $\mathcal{Q}$ opts $\mathbf{z} \in Z_n$ as their mixed strategies then the expected payoff for Player $\mathcal{P}$ is determined as

$$E_P(\mathbf{y}, \mathbf{z}) = \oplus_{j=1}^{m} \oplus_{k=1}^{n} y_j \tilde{\tilde{\Gamma}}_{jk} z_k$$
$$= \oplus_{j=1}^{m} \oplus_{k=1}^{n} y_j \left\langle \left(\tau_{p\Gamma}^{jk}, \tau_{s\Gamma}^{jk} \right); \left(\iota_{p\Gamma}^{jk}, \iota_{s\Gamma}^{jk} \right); \left(\xi_{p\Gamma}^{jk}, \xi_{s\Gamma}^{jk} \right) \right\rangle z_k. \qquad (4)$$

We can calculate the expected payoff for Player $\mathcal{Q}$ in similar way as

$$E_Q(\mathbf{y}, \mathbf{z}) = \oplus_{j=1}^{m} \oplus_{k=1}^{n} y_j \tilde{\tilde{\Upsilon}}_{jk} z_k$$
$$= \oplus_{j=1}^{m} \oplus_{k=1}^{n} y_j \left\langle \left(\tau_{p\Upsilon}^{jk}, \tau_{s\Upsilon}^{jk} \right); \left(\iota_{p\Upsilon}^{jk}, \iota_{s\Upsilon}^{jk} \right); \left(\xi_{p\Upsilon}^{jk}, \xi_{s\Upsilon}^{jk} \right) \right\rangle z_k. \qquad (5)$$

Now, we define Nash equilibrium solution (NES) of bimatrix games in neutrosophic environment as follows.

Definition 5 A pair $(\mathbf{y}^*, \mathbf{z}^*) \in (Y_m, Z_n)$ is called a NES of the neutrosophic bimatrix game $\tilde{\tilde{B}}$ if the pair $(\mathbf{y}^*, \mathbf{z}^*)$ satisfies the undermentioned inequalities

$$\mathbf{y}^T \tilde{\tilde{\Gamma}} \mathbf{z}^* \leq \mathbf{y}^{*T} \tilde{\tilde{\Gamma}} \mathbf{z}^*, \ \forall \mathbf{y} \in Y_m \text{ and } \mathbf{y}^{*T} \tilde{\tilde{\Upsilon}} \mathbf{z} \leq \mathbf{y}^{*T} \tilde{\tilde{\Upsilon}} \mathbf{z}^*, \ \forall \mathbf{z} \in Z_n.$$

Such a pair $(\mathbf{y}^*, \mathbf{z}^*)$ is called the optimal solution of the bimatrix game $\tilde{\tilde{B}}$. Moreover, $\mathbf{y}^{*T} \tilde{\tilde{\Gamma}} \mathbf{z}^*$ and $\mathbf{y}^{*T} \tilde{\tilde{\Upsilon}} \mathbf{z}^*$ are termed as the optimal values of the bimatrix game $\tilde{\tilde{B}}$ for Players $\mathcal{P}$ and $\mathcal{Q}$, respectively.

In 1995, Owen [16] proved that every bimatrix game approaches at least one equilibrium solution. Extending the thought of Owen, we can say that determining the NES of a neutrosophic bimatrix game is equivalent to resolve the following quadratic type-2 neutrosophic fuzzy programming problem (QT2NFPP):

$$\max \ \{\mathbf{y}(\tilde{\tilde{\Gamma}} + \tilde{\tilde{\Upsilon}})\mathbf{z}\} - \tilde{\tilde{V}}_P - \tilde{\tilde{V}}_Q$$

$$\text{subject to} \ \ \mathbf{y}\tilde{\tilde{\Gamma}} \le \tilde{\tilde{V}}_P e_m, \ \tilde{\tilde{\Upsilon}}\mathbf{z} \le \tilde{\tilde{V}}_Q e_n \tag{6}$$

$$\mathbf{y} \in Y_m, \mathbf{z} \in Z_n.$$

Further, if $(\mathbf{y}^*, \mathbf{z}^*, \tilde{\tilde{V}}_P^*, \tilde{\tilde{V}}_Q^*)$ be the NES of the bimatrix game $\tilde{\tilde{B}}$, then we have $\tilde{\tilde{V}}_P^* = \mathbf{y}^* \tilde{\tilde{\Gamma}} \mathbf{z}^*$, $\tilde{\tilde{V}}_Q^* = \mathbf{y}^* \tilde{\tilde{\Upsilon}} \mathbf{z}^*$ and $\mathbf{y}^{*T} (\tilde{\tilde{\Gamma}} + \tilde{\tilde{\Upsilon}})\mathbf{z}^* - \tilde{\tilde{V}}_P^* - \tilde{\tilde{V}}_Q^* = 0$.

We can rewrite Problem (6) as

$$\max \ \{\oplus_{j=1}^m \oplus_{k=1}^n y_j(\tilde{\tilde{\Gamma}}_{jk} + \tilde{\tilde{\Upsilon}}_{jk})z_k\} - \tilde{\tilde{V}}_P - \tilde{\tilde{V}}_Q$$

$$\text{subject to} \ \ \oplus_{j=1}^m y_j \tilde{\tilde{\Gamma}}_{jk} \le \tilde{\tilde{V}}_P \ \text{for } k = 1, 2, \dots, n$$

$$\oplus_{k=1}^n \tilde{\tilde{\Upsilon}} z_k \le \tilde{\tilde{V}}_Q \ \text{for } j = 1, 2, \dots, m \tag{7}$$

$$\sum_{j=1}^m y_j, \sum_{k=1}^n z_k = 1; \ y_j, z_k \ge 0 \ \text{for } j = 1, 2, \dots, m; k = 1, 2, \dots, n.$$

It is difficult to resolve Problem (7) in traditional approach. Thus, to evaluate the NES of the bimatrix game $\tilde{\tilde{B}}$, we extend the notion of the composite relative degree of similarity index to the PIS [4] in the T2NF scenario. We establish the present methodology based on the thought that the players' best choice should have the greatest distance from the positive ideal type-2 neutrosophic fuzzy solution (PIT2NFS). Also, players' best choice should have the lowest distance from the negative ideal type-2 neutrosophic fuzzy solution (NIT2NFS). To determine the NES, we have to follow the under-mentioned way.

Step 1: Calculate PIT2NFS and NIT2NFS for the columns of the payoff matrix $\tilde{\tilde{\Gamma}}$ as

$$\tilde{\tilde{\Gamma}}_j^+ = \max_{1 \le j \le m} \tilde{\tilde{\Gamma}}_{jk} = \left\langle \left(\max_{1 \le j \le m} \tau_{p\Gamma}^{jk}, \max_{1 \le j \le m} \tau_{s\Gamma}^{jk} \right); \left(\min_{1 \le j \le m} \iota_{p\Gamma}^{jk}, \min_{1 \le j \le m} \iota_{s\Gamma}^{jk} \right);$$

$$\left(\min_{1 \le j \le m} \xi_{p\Gamma}^{jk}, \min_{1 \le j \le m} \xi_{s\Gamma}^{jk} \right) \right\rangle \tag{8}$$

$$\tilde{\tilde{\Gamma}}_j^- = \min_{1 \le j \le m} \tilde{\tilde{\Gamma}}_{jk} = \left\langle \left(\min_{1 \le j \le m} \tau_{p\Gamma}^{jk}, \min_{1 \le j \le m} \tau_{s\Gamma}^{jk} \right); \left(\max_{1 \le j \le m} \iota_{p\Gamma}^{jk}, \max_{1 \le j \le m} \iota_{s\Gamma}^{jk} \right);$$

$$\left(\max_{1 \le j \le m} \xi_{p\Gamma}^{jk}, \max_{1 \le j \le m} \xi_{s\Gamma}^{jk} \right) \right\rangle \tag{9}$$

Step 2: Using Eq. (3), evaluate the similarity index $S_e(\tilde{\tilde{\Gamma}}_j^+, \tilde{\tilde{\Gamma}}_{jk})$ and $S_e(\tilde{\tilde{\Gamma}}_j^-, \tilde{\tilde{\Gamma}}_{jk})$ of the T2NF entry $\tilde{\tilde{\Gamma}}_{jk}$ to the PIT2NFS $\tilde{\tilde{\Gamma}}_j^+$ and the NIT2NFS $\tilde{\tilde{\Gamma}}_j^-$, respectively, for $j = 1, 2, \ldots, m$.

Step 3: Find the relative index $\mathcal{R}_{jk}$ using the rule

$$\mathcal{R}_{jk} = \frac{S_e(\tilde{\tilde{\Gamma}}_j^+, \tilde{\tilde{\Gamma}}_{jk})}{S_e(\tilde{\tilde{\Gamma}}_j^+, \tilde{\tilde{\Gamma}}_{jk}) + S_e(\tilde{\tilde{\Gamma}}_j^-, \tilde{\tilde{\Gamma}}_{jk})},$$

for $j = 1, 2, \ldots, m$. Also, shape the relative similarity matrix $\Gamma^* = (\mathcal{R}_{jk})_{m \times n}$ for Player $\mathcal{P}$. Γ^* is taken as the crisp equivalent of the payoff matrix $\tilde{\tilde{\Gamma}}$.

Step 4: Determine PIT2NFS and NIT2NFS for the rows of the payoff matrix $\tilde{\tilde{\Upsilon}}$ as

$$\tilde{\tilde{\Upsilon}}_k^+ = \max_{1 \le k \le n} \tilde{\tilde{\Upsilon}}_{jk} = \Big\langle \Big(\max_{1 \le k \le n} \tau_{p\Upsilon}^{jk}, \max_{1 \le k \le n} \tau_{s\Upsilon}^{jk} \Big); \Big(\min_{1 \le k \le n} \iota_{p\Upsilon}^{jk}, \min_{1 \le k \le n} \iota_{s\Upsilon}^{jk} \Big);$$
$$\Big(\min_{1 \le k \le n} \xi_{p\Upsilon}^{jk}, \min_{1 \le k \le n} \xi_{s\Upsilon}^{jk} \Big) \Big\rangle \tag{10}$$

$$\tilde{\tilde{\Upsilon}}_k^- = \min_{1 \le k \le n} \tilde{\tilde{\Upsilon}}_{jk} = \Big\langle \Big(\min_{1 \le k \le n} \tau_{p\Upsilon}^{jk}, \min_{1 \le k \le n} \tau_{s\Upsilon}^{jk} \Big); \Big(\max_{1 \le k \le n} \iota_{p\Upsilon}^{jk}, \max_{1 \le j \le m} \iota_{s\Upsilon}^{jk} \Big);$$
$$\Big(\max_{1 \le j \le m} \xi_{p\Upsilon}^{jk}, \max_{1 \le k \le n} \xi_{s\Upsilon}^{jk} \Big) \Big\rangle \tag{11}$$

Step 5: Similar to **Step 2**, calculate the similarity index $S_e(\tilde{\tilde{\Upsilon}}_k^+, \tilde{\tilde{\Upsilon}}_{jk})$ and $S_e(\tilde{\tilde{\Upsilon}}_k^-, \tilde{\tilde{\Upsilon}}_{jk})$ of the T2NF entry $\tilde{\tilde{\Upsilon}}_{jk}$ to the PIT2NFS $\tilde{\tilde{\Upsilon}}_k^+$ and the NIT2NFS $\tilde{\tilde{\Upsilon}}_k^-$, respectively, for $k = 1, 2, \ldots, n$, using Eq. (3).

Step 6: Determine the relative similarity index $\mathcal{R}'_{jk}$ using

$$\mathcal{R}'_{jk} = \frac{S_e(\tilde{\tilde{\Upsilon}}_k^+, \tilde{\tilde{\Upsilon}}_{jk})}{S_e(\tilde{\tilde{\Upsilon}}_k^+, \tilde{\tilde{\Upsilon}}_{jk}) + S_e(\tilde{\tilde{\Upsilon}}_k^-, \tilde{\tilde{\Upsilon}}_{jk})},$$

for $k = 1, 2, \ldots, n$. Now, construct the relative similarity matrix $\Upsilon^* = (\mathcal{R}'_{jk})_{m \times n}$ for Player $\mathcal{Q}$. Note that, Υ^* is the crisp equivalent of the payoff matrix $\tilde{\tilde{\Upsilon}}$.

Step 7: Thus, we transform QT2NFPP (7) into the following

$$\max\{\sum_{j=1}^{m}\sum_{k=1}^{n}y_j(\mathcal{R}_{jk}+\mathcal{R}'_{jk})z_k\} - v_P - v_Q$$

subject to $\sum_{j=1}^{m}y_j\mathcal{R}_{jk} \leq v_P$ for $k = 1, 2, \ldots, n$

$$\sum_{k=1}^{n}\mathcal{R}'_{jk}z_k \leq v_Q \text{ for } j = 1, 2, \ldots, m \tag{12}$$

$$\sum_{j=1}^{m}y_j, \sum_{k=1}^{n}z_k = 1; \ y_j, z_k \geq 0 \text{ for } j = 1, 2, \ldots, m; k = 1, 2, \ldots, n.$$

Here, v_P and v_Q stand for the crisp equivalents of $\tilde{\tilde{V}}_P$ and $\tilde{\tilde{V}}_Q$, respectively.

Problem (12) is the crisp equivalent of Problem (7). Now, solving Problem (12), we can obtain the optimal strategies $\mathbf{y}^*$ and $\mathbf{z}^*$ for Players $\mathcal{P}$ and $\mathcal{Q}$, respectively, along with the optimal values of the game v_P and v_Q.

6 Numerical Illustration

To justify the proposed methodology and to check its applicability in real-world situation we discuss the following problem.

6.1 An Application to Mitigate Environmental Pollution

Due to rapid industrialization, unplanned urbanization, etc., environmental pollution increases daily. Regarding environmental pollution management issues, the government and some non-governmental organizations (NGOs) adopt a few strategies as the two players in the game. Both the government and the NGOs aim to maximize citizens' awareness to mitigate the pollution level. Suppose government takes two strategies, a_1: positive supervision of every action regarding environmental pollution, and a_2: implement numerous regulations to maintain the status quo. Similarly, NGOs take some strategies; likewise, b_1: conduct multiple ecological awareness programs, and b_2: arrange some 'go-green activities'. However, the government and the NGOs take whatever strategies; their implementation depends upon a few parameters, such as the adaptability of local people, their daily needs, etc. So, to figure out the amount of reduction of pollution level, a 2-dimensional fuzzy set is not enough. Thus, we portray such a situation through a bimatrix game problem with payoffs of T2NFSs.

Suppose the government and the NGOs are assigned as Players $\mathcal{P}$ and $\mathcal{Q}$, respectively. Now, if Player $\mathcal{P}$ chooses strategy $a_j (j = 1, 2)$ and $\mathcal{Q}$ opts strategy

Table 1 T2NFS entries of the payoff matrices

Entries	Linguistic terms	T2NFS values
$\tilde{\tilde{\Gamma}}_{11}$	High	$\{\langle(0.8, 0.9); (0.2, 0.1); (0.3, 0.4)\rangle, \langle(0.7, 0.7); (0.1, 0.3); (0.1, 0.2)\rangle\}$
$\tilde{\tilde{\Gamma}}_{12}$	Average	$\{\langle(0.5, 0.5); (0.2, 0.1); (0.2, 0.1)\rangle, \langle(0.6, 0.5); (0.3, 0.2); (0.1, 0.2)\rangle\}$
$\tilde{\tilde{\Gamma}}_{21}$	Low	$\{\langle(0.3, 0.4); (0.3, 0.2); (0.2, 0.2)\rangle, \langle(0.5, 0.4); (0.2, 0.1); (0.1, 0.4)\rangle\}$
$\tilde{\tilde{\Gamma}}_{22}$	High	$\{\langle(0.9, 0.7); (0.3, 0.5); (0.2, 0.1)\rangle, \langle(0.6, 0.7); (0.2, 0.2); (0.1, 0.2)\rangle\}$
$\tilde{\tilde{\Upsilon}}_{11}$	High	$\{\langle(0.9, 0.8); (0.6, 0.5); (0.2, 0.2)\rangle, \langle(0.7, 0.8); (0.1, 0.3); (0.1, 0.3)\rangle\}$
$\tilde{\tilde{\Upsilon}}_{12}$	Average	$\{\langle(0.6, 0.5); (0.2, 0.4); (0.1, 0.1)\rangle, \langle(0.6, 0.7); (0.1, 0.1); (0.2, 0.2)\rangle\}$
$\tilde{\tilde{\Upsilon}}_{21}$	Low	$\{\langle(0.3, 0.5); (0.2, 0.3); (0.1, 0.1)\rangle, \langle(0.4, 0.4); (0.1, 0.2); (0.2, 0.2)\rangle\}$
$\tilde{\tilde{\Upsilon}}_{22}$	High	$\{\langle(0.8, 0.9); (0.2, 0.1); (0.3, 0.5)\rangle, \langle(0.6, 0.7); (0.3, 0.2); (0.1, 0.2)\rangle\}$

$b_k (k = 1, 2)$, the payoff matrices for $\mathcal{P}$ and $\mathcal{Q}$, respectively are assigned with some linguistic terms as $\tilde{\tilde{\Gamma}} = (\tilde{\tilde{\Gamma}}_{jk})_{2\times2} = \begin{pmatrix} High & Average \\ Low & High \end{pmatrix}$ and $\tilde{\tilde{\Upsilon}} = (\tilde{\tilde{\Upsilon}}_{jk})_{2\times2} = \begin{pmatrix} High & Average \\ Low & High \end{pmatrix}$. Linguistic entries of the payoff matrices and their T2NFS representations are enlisted in Table 1.

Here, the entries of the payoff matrices signifies the amount of reduction of the pollution level yearly. Now, we can solve the problem using few steps as depicted in Sect. 5.

Step 6.1: Using Eqs. (8) and (9) calculate PIT2NFS and NIT2NFS, respectively for Player $\mathcal{P}$ as

$$\tilde{\tilde{\Gamma}}_1^+ = \{\langle(0.8, 0.9); (0.2, 0.1); (0.2, 0.2)\rangle, \langle(0.7, 0.7); (0.1, 0.1); (0.1, 0.2)\rangle\}$$

$$\tilde{\tilde{\Gamma}}_2^+ = \{\langle(0.9, 0.7); (0.2, 0.1); (0.2, 0.1)\rangle, \langle(0.6, 0.7); (0.2, 0.2); (0.1, 0.2)\rangle\}$$

$$\tilde{\tilde{\Gamma}}_1^- = \{\langle(0.3, 0.4); (0.3, 0.2); (0.3, 0.4)\rangle, \langle(0.5, 0.4); (0.2, 0.3); (0.1, 0.4)\rangle\}$$

$$\tilde{\tilde{\Gamma}}_2^- = \{\langle(0.5, 0.5); (0.3, 0.5); (0.2, 0.1)\rangle, \langle(0.6, 0.5); (0.3, 0.2); (0.1, 0.2)\rangle\}$$

Step 6.2: Utilizing Eq. (3), we evaluate the similarity index of the T2NFS entries to the PIT2NFS and NIT2NFS.

$$S_e(\tilde{\tilde{\Gamma}}_1^+, \tilde{\tilde{\Gamma}}_{11}) = 0.7652, \quad S_e(\tilde{\tilde{\Gamma}}_1^+, \tilde{\tilde{\Gamma}}_{21}) = 0.8779,$$

$$S_e(\tilde{\tilde{\Gamma}}_2^+, \tilde{\tilde{\Gamma}}_{12}) = 0.9361, \quad S_e(\tilde{\tilde{\Gamma}}_2^+, \tilde{\tilde{\Gamma}}_{22}) = 0.9474,$$

$$S_e(\tilde{\tilde{\Gamma}}_1^-, \tilde{\tilde{\Gamma}}_{11}) = 0.8779, \quad S_e(\tilde{\tilde{\Gamma}}_1^-, \tilde{\tilde{\Gamma}}_{21}) = 0.7652,$$

$$S_e(\tilde{\tilde{\Gamma}}_2^-, \tilde{\tilde{\Gamma}}_{12}) = 0.9361, \quad S_e(\tilde{\tilde{\Gamma}}_2^-, \tilde{\tilde{\Gamma}}_{22}) = 0.9361.$$

Step 6.3: Construct relative similarity matrix Γ^* for Player $\mathcal{P}$ as

$$\Gamma^* = (\mathcal{R}_{jk})_{2\times 2} = \begin{pmatrix} 0.4657 & 0.5 \\ 0.5343 & 0.5029 \end{pmatrix}.$$

Step 6.4: Again utilising Eqs. (8) and (9) determine PIT2NFS and NIT2NFS, respectively for Player $\mathcal{Q}$ as

$$\tilde{\tilde{\Upsilon}}_1^+ = \{\langle(0.9, 0.8); (0.2, 0.4); (0.1, 0.1)\rangle, \langle(0.7, 0.8); (0.1, 0.1); (0.1, 0.2)\rangle\}$$

$$\tilde{\tilde{\Upsilon}}_2^+ = \{\langle(0.8, 0.9); (0.2, 0.1); (0.1, 0.1)\rangle, \langle(0.6, 0.7); (0.1, 0.2); (0.1, 0.2)\rangle\}$$

$$\tilde{\tilde{\Upsilon}}_1^- = \{\langle(0.6, 0.5); (0.6, 0.5); (0.2, 0.2)\rangle, \langle(0.6, 0.7); (0.1, 0.3); (0.2, 0.3)\rangle\}$$

$$\tilde{\tilde{\Upsilon}}_2^- = \{\langle(0.3, 0.5); (0.2, 0.3); (0.3, 0.5)\rangle, \langle(0.4, 0.4); (0.3, 0.2); (0.2, 0.2)\rangle\}$$

Step 6.5: Utilizing Eq. (3), we evaluate the similarity index of the T2NFS entries to the PIT2NFS and NIT2NFS for Player $\mathcal{Q}$.

$$S_e(\tilde{\tilde{\Gamma}}_1^+, \tilde{\tilde{\Gamma}}_{11}) = 0.9292, \quad S_e(\tilde{\tilde{\Gamma}}_1^+, \tilde{\tilde{\Gamma}}_{12}) = 0.9641,$$

$$S_e(\tilde{\tilde{\Gamma}}_2^+, \tilde{\tilde{\Gamma}}_{21}) = 0.8868, \quad S_e(\tilde{\tilde{\Gamma}}_2^+, \tilde{\tilde{\Gamma}}_{22}) = 0.9355,$$

$$S_e(\tilde{\tilde{\Gamma}}_1^-, \tilde{\tilde{\Gamma}}_{11}) = 0.8838, \quad S_e(\tilde{\tilde{\Gamma}}_1^-, \tilde{\tilde{\Gamma}}_{12}) = 0.9292,$$

$$S_e(\tilde{\tilde{\Gamma}}_2^-, \tilde{\tilde{\Gamma}}_{21}) = 0.9355, \quad S_e(\tilde{\tilde{\Gamma}}_2^-, \tilde{\tilde{\Gamma}}_{22}) = 0.8927.$$

Step 6.6: Form the relative similarity matrix Υ^* for Player $\mathcal{Q}$ as

$$\Upsilon^* = (\mathcal{R}'_{jk})_{2\times 2} = \begin{pmatrix} 0.5125 & 0.5092 \\ 0.4866 & 0.5117 \end{pmatrix}.$$

Table 2 Results for a bimatrix games with T2NFS entries

Player	Optimal strategy	Value of game	Expected payoff
$\mathcal{P}$	$\mathbf{y}^* = (0.4779, 0.5221)$	$v_P = 0.5015$	$\tilde{\tilde{E}}_P(\mathbf{y}^*, \mathbf{z}^*) =$ $\{\langle(0.63, 0.46); (0.24, 0.15);$ $(0.20, 0.11)\rangle, \langle(0.43, 0.44);$ $(0.23, 0.20); (0.10, 0.21)\rangle$
$\mathcal{Q}$	$\mathbf{z}^* = (0.0880, 0.9120)$	$v_P = 0.5095$	$\tilde{\tilde{E}}_Q(\mathbf{y}^*, \mathbf{z}^*) =$ $\{\langle(0.56, 0.63); (0.21, 0.21);$ $(0.17, 0.22)\rangle, \langle(0.43, 0.53);$ $(0.17, 0.15); (0.14, 0.20)\rangle$

Step 6.7: Using the values of $\mathcal{R}_{jk}$ and $\mathcal{R}'_{jk}$ in Problem (12), we have

$$\max\{0.9782y_1z_1 + 1.0092y_1z_2 + 1.02009y_2z_1 + 1.0146y_2z_2\} - v_P - v_Q$$
$$\text{subject to } 0.4657y_1 + 0.5343y_2 \leq v_P$$
$$0.5y_1 + 0.5029y_2 \leq v_P$$
$$0.5125z_1 + 0.5092z_2 \leq v_Q$$
$$0.4866z_1 + 0.5117z_2 \leq v_Q$$
$$\sum_{j=1}^{2} y_j, \sum_{k=1}^{2} z_k = 1; \ y_j, z_k \geq 0 \text{ for } j = 1, 2; k = 1, 2. \tag{13}$$

Solving Problem (13), we obtain $(\mathbf{y}^*, \mathbf{z}^*)$, the optimal strategies, the value of the games v_P, v_Q for Players $\mathcal{P}$ and $\mathcal{Q}$, respectively. We gather the obtained NES in Table 2. Further, substituting the values of optimal strategies $\mathbf{y}^*$ and $\mathbf{z}^*$ in Eqs. (4) and (5) we determine the expected payoffs for Players $\mathcal{P}$ and $\mathcal{Q}$, which is depicted in Table 2.

From Table 2, we observe that the expected payoff for Player $\mathcal{P}$ is $\tilde{\tilde{E}}_P(\mathbf{y}^*, \mathbf{z}^*) = \{\langle(0.63, 0.46); (0.24, 0.15); (0.20, 0.11)\rangle, \langle(0.43, 0.44); (0.23, 0.20); (0.10, 0.21)\rangle$ and for Player $\mathcal{Q}$ is $\tilde{\tilde{E}}_Q(\mathbf{y}^*, \mathbf{z}^*) = \{\langle(0.56, 0.63); (0.21, 0.21); (0.17, 0.22)\rangle, \langle(0.43, 0.53); (0.17, 0.15); (0.14, 0.20)\rangle$. This indicates if the government implement strategy a_1 with probability 0.4779, strategy a_2 with probability 0.5221 and NGOs implement strategy b_1 with probability 0.0880, strategy b_2 with probability 0.9120, then the amount of reducing the pollution will be 'average' according to the estimation of T2NF linguistic variable depicted in Table 1.

7 Conclusion

Aiming to consider the truth, indeterminacy, and falsity of membership value of the large volume of data, having multiple sources, we have presented a bimatrix game in the T2NF background. To determine the solution to such a problem, we offer a similarity index of T2NFSs, built up based o the normalized Euclidean distance of T2NFSs. It was a challenging task to handle T2NFS payoffs. Suppose we want to solve such a T2NF bimatrix game in the traditional method. In that case, we need a defuzzificaion function, and the use of that defuzzification function loses the intake data by a wide margin. So, to avoid such irregularity, we have solved the bimatrix game using players' best choices' composite relative degree to the positive ideal solution in the T2NF scenario. It is an excellent contribution to the present study.

In the present discussion, we have assumed that there must exist a solution to the T2NF bimatrix game. But, we could not prove this assurance. This is a drawback of the present study. To overcome such a fault, we need further investigation. Besides this, the researcher may imply the rendered model to explain various market management problems, water-supply management problems, military science, urban management problems, etc.

References

1. An, J.J., Li, D.F., Nan, J.X.: A mean-area ranking based non-linear programming approach to solve intuitionistic fuzzy bi-matrix games. J. Intell. Fuzzy Syst. **33**(1), 563–573 (2017)
2. Bhaumik, A., Roy, S.K., Li, D.F.: (α, β, γ)—cut set based ranking approach to solving bi-matrix games in neutrosophic environment. Soft Comput. (2020). https://doi.org/10.1007/s00500-020-05332-6
3. Brikaa, M.G., Ammer, E.S., Zhang, Z.: Solving bi-matrix games in tourism planning management under rough interval approach. Int. J. Math. Sci. Comput. **4**, 44–62 (2019)
4. Jana, J., Roy, S.K.: Dual hesitant fuzzy matrix games: based on new similarity measure. Soft Comput. **23**(1), 8873–8886 (2019)
5. Karaaslan, F., Hunu, F.: Type-2 single-valued neutrosophic sets and their applications in multi-criteria group decision making based on TOPSIS method. J. Amb. Intell. Human. Comput. **11**, 4113–4132 (2020)
6. Karmakar S., Seikh M. R., Castillo O.: Type-2 intuitionistic fuzzy matrix games based on a new distance measure: application to biogas-plant implementation problem. Appl. Soft Comput. **106** (2021). https://doi.org/10.1016/j.asoc.2021:107357
7. Karmakar S., Seikh M.R.: A novel ranking-based non-linear programming approach to solve bimatrix games in dense fuzzy environment. In: Proceeding of the Seventh International Conference on Mathematics and Computing. Advance in Intelligent Systems and Computing, vol. 1412 (2022). https://doi.org/10.1007/978-981166890-6_56
8. Khan, I., Mehra, A.: A novel equilibrium solution concept for intuitionistic fuzzy bi-matrix games considering proportion mix of possibility and necessity expectations. Gran. Comput. **5**, 461–483 (2020)
9. Li, C., Li, M.: A new bi-matrix game model with fuzzy pay-offs in credibility space. Int. J. Fuzzy Comput. Model. **10**(5), 556–563 (2019)

10. Li, D.F., Yang, J.: A difference-index based ranking bilinear programming approach to solving bi-matrix games with pay-offs of trapezoidal intuitionistic fuzzy numbers. J. Appl. Math. **13**, 1–13 (2013)
11. Liu, K., Xing, Y.: Solving bi-matrix games based on fuzzy payoffs via utilizing the interval value function method. Mathematics **7**(5), 469 (2019). https://doi.org/10.3390/math7050469
12. Li, S., Tu, G.: Bi-matrix games with general intuitionistic fuzzy payoffs and application in corporate environmental behavior. Soft Comput. **14**, 671 (2022). https://doi.org/10.3390/sym14040671
13. Mendel, J.M.: Uncertain Rule-based Fuzzy Logic System: Introduction and New Direction. Prentice-Hall, New Jersey (2001)
14. Nan, J.X., Li, D.F., An, J.J.: Solving bi-matrix games with intuitionistic fuzzy pay-offs. J. Intell. Fuzzy Syst. **33**(6), 3723–3732 (2017)
15. Nash, J.E.: Non cooperative games. Ann. Math. **54**(2), 286–295 (1951)
16. Owen, G.: Game Theory, 3rd edn. Academic Press, New York (1995)
17. Ozlu, S., Karaaslan, F.: Hybrid similarity measures of single-valued type-2 neutrosophic sets and their application to MCDM based TOPSIS. Soft Comput. (2022). https://doi.org/10.1007/s00500-022-06824-3
18. Roy, S.K., Bhaumik, A.: Intelligent water management: a triangular type-2 intuitionistic fuzzy matrix game approach. Water Resourc. Manag. **32**, 949–968 (2018)
19. Roy S.K., Maiti S.K.: Reduction methods of type-2 fuzzy variables and their application to stackelberg game. Appl. Intell. **50**, 1398–1415 (2020)
20. Seikh, M.R., Nayak, P.K., Pal, M.: Solving bi-matrix games with pay-offs of triangular intuitionistic fuzzy numbers. Eur. J. Pure Appl. Math. **8**(2), 153–171 (2015)
21. Seikh, M.R., Nayak, P.K., Pal, M.: An alternative approach to solve bi-matrix games with intuitionistic fuzzy goals. Int. J. Fuzzy Comput. Model. **1**(4), 362–381 (2016)
22. Seikh, M.R., Karmakar, S., Castillo, O.: A novel defuzzification approach of type-2 fuzzy variable to solving matrix games: an application to plastic ban problem. Iran. J. Fuzzy Syst. **18**(5), 155–172 (2021)
23. Smarandache F.: First International Conference on Neutrosophy, Neutrosophic Logic, Set, Probability And Statistics. Florentin (2001)
24. Yang, J., Fei, W., Li, D.F.: Non-linear programming approach to solve bi-matrix games with payoffs represented by I-fuzzy numbers. Int. J. Fuzzy Syst. **18**, 492–503 (2016)
25. Zhang, W., Xing, Y., Qiu, D.: Multiobjective fuzzy bi-matrix game model: a multicriteria non-linear programming approach. Symmetry **9**(8), 159 (2017). https://doi.org/10.3390/sym9080159

Discouraged Arrivals Queueing System in Interval Valued Type-2 Fuzzy Environment

R. Ramesh and M. Seenivasan

Abstract We provide a survey in this article about cost and attainment stratagems for discouraged arrivals Queueing System in the interval valued type-2 fuzzy environment by centroid of centroids ordering technique. This technique is very remarkable for defuzzification and it helps to convert the fuzzy numerals to crisp (classic) numerals. This propounded ordering technique is very effective and provides the accurate verdicts.

Keywords Type 1 fuzzy sets · Interval-Valued Type-2 Fuzzy Sets (IVT2FS) · Discouraged arrivals queueing system · Centroid of centroids ordering · Costs

1 Introduction

Queueing theory occupies a prominent place in our daily life environment. Mainly we meet massive queues in front of Temples, Hotels, Mobile networks, ATM, Theaters, Hospitals, Malls, etc....,. The ultimate aim of the system manager is the Time governance. Here the queueing postulates employs a tremendous role.

Normal approaches may not provide a new aspect of research. So every day we have to update with some new ideas. Here we show a new path for queueing models. Queueing theory concepts and models can be applied in many fuzzy environments. More authors have been approached with type 1 fuzzy sets and systems in queueing models. Already we also analyzed the queueing theory models in type 1 fuzzy environment. Now we are analyzing the Discouraged arrivals queueing models in type 2 fuzzy environments, particularly in interval-valued type-2 fuzzy sets (IVT2FS). It can give different outlooks and furnishes more precisions. This approach can

R. Ramesh
Department of Mathematics, Arignar Anna Government Arts College, Musiri, Tamil Nadu, India

M. Seenivasan (✉)
Mathematics Wing, DDE, Annamalai University, Annamalainagar, Tamil Nadu, India
e-mail: emseeni@yahoo.com

provide extraordinary perceptions instead of typical access. This is a new insight and development in the research field.

The queueing preludes [5, 11] are very useful to solve the congesting problems. We can solve these problems by fuzzy manner [12]. A long year ago Natvig analyzed the discouraged arrivals queueing system. Then some authors discussed with discouraged arrivals queues [3, 19]. The retention queuing system [13] surveyed by Kumar and Sharma.

In the earlier period, Prof. Zadeh introduced and extended the fuzzy or type-1 fuzzy sets (T1FS). After that Mizumoto introduced some properties of (T2FS). Then (IVT2FS) was extended by Mendel et al. [14, 15]. Recently Alterio et al. [2] and Shuwang Qin et al. [18] analyzed the (IVT2FS) and logic.

The Type-2 fuzzy systems are the generalization of Type-1 fuzzy systems and have more uncertainty states. Type1 fuzzy systems are functioning with a specified membership function, but the membership function of type-2 fuzzy systems is varying.

An (IVT2FS) is occasionally determined by first-order and a general (T2FS) is periodically determined by second-order model of uncertainty fuzzy set. It is feasible to apply the (IVT2FS) instead of type-1 fuzzy sets.

More inventors utilized sundry ordering methods for defuzzifying the fuzzy numerals. Some researchers are used distance and area rooted ordering techniques [1, 4, 6, 7, 17, 20, 21]. Some inventors used centroid grounded ordering methods [10, 22, 23] and Wingspans ordering technique [24]. Also some of the authors [8, 9, 16, 25] presented more ranking techniques for interval-valued type-2 Fuzzy numbers (IVT2FN). Our provided ordering procedure is too better than the other.

2 Preludes

Definition 1 A (T1FS), $\tilde{A} = \{(x, \varphi_{\tilde{A}}(x)); x \in X\}$ is indented by a membership function $\varphi_{\tilde{A}}$ and a mapping $\varphi_{\tilde{A}}: X \to [0, 1]$ where X is the universal set.

Definition 2 A Fuzzy Set $\tilde{A}$ is called (T2FS) if it is represented by
$\tilde{A} = \{[(x,u), \varphi_{\tilde{A}}(x,u)]; \forall x \in X, \forall u \in J_x \subseteq [0,1]\}, 0 \le \varphi_{\tilde{A}}(x,u) \le 1$. Here J_x is primary and $\varphi_{\tilde{A}}(x,u)$ is secondary membership functions of x. This secondary membership function produces the chances over the primary membership function.

Definition 3 A fuzzy set $\tilde{A}$ is said to be (IVT2FS), if it has an upper and lower membership functions respectively $\phi_{\tilde{A}}^{U}(x) \in [0,1], \phi_{\tilde{A}}^{L}(x) \in [0,1]$ is termed by
$\tilde{A} = \{[\phi_{\tilde{A}}^{U}(x), \phi_{\tilde{A}}^{L}(x)]; \forall x \in U\}$. This is a special case of (T2FS), whose all the secondary membership values are equal to 1.

Definition 4 A type-1 Trapezoidal Fuzzy Number (TrFN) $\tilde{A}$ (a,b,c,d;1) should have the membership function

Fig. 1 Type-1 TrFN

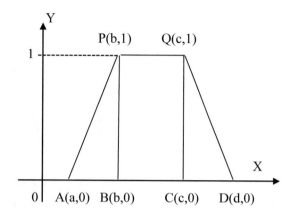

$$\phi_{\tilde{A}}(x) = \begin{cases} \frac{x-a}{b-a}, & a \le x \le b \\ 1, & b \le x \le c \\ \frac{x-d}{c-d}, & c \le x \le d \\ 0, & otherwise \end{cases}$$ and the diagrammatic explanation is given in

Fig. 1.

Definition 5 An (IVT2FN) $\tilde{A} = [A^U, A^L] = [(a^U, b^U, c^U, d^U; h_1)(a^L, b^L, c^L, d^L; h_2)]$ should have the upper membership function.

$$\phi_{\tilde{A}}{}^{U}(x) = \begin{cases} \frac{h_1(x-a^U)}{b^U-a^U}, & a^U \le x \le b^U \\ h_1, & b^U \le x \le c^U \\ \frac{h_1(x-d^U)}{c^U-d^U}, & c^U \le x \le d^U \\ 0, & otherwise \end{cases}$$

and the lower membership function

$$\phi_{\tilde{A}}{}^{L}(x) = \begin{cases} \frac{h_2(x-a^L)}{b^L-a^L}, & a^L \le x \le b^L \\ h_2, & b^L \le x \le c^L \\ \frac{h_2(x-d^L)}{c^U-d^U}, & c^L \le x \le d^L \\ 0, & otherwise \end{cases}$$

here h_1 and h_2 are the upper height and lower height of the membership functions and the diagrammatic explanation is given in Fig. 2.

Fig. 2 IVT2FN

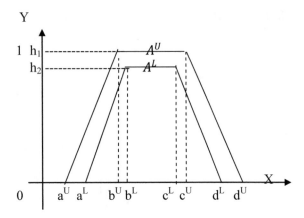

3 Queue Formation

Think about the customers appearing along Poisson structure in a sole server with the fuzzy rate $\tilde{\lambda}$. They served by exponentially with fuzzy rate $\tilde{\mu}$ in FIFO stream. If the system grows up, then the arrival rate decreases moderately. The new arrival (Fig. 3) gets discourage (discouraged arrivals) to enter the line.

This figure expresses that a woman discourages her entry in the ration shop queue due to the long queue juncture.

Fig. 3 Discouraged arrivals

(I) **Performance Expedients**

We take the finite capacity of the system as N. Each patron can stand by in the queue in a specific service time duration (T). In the end of T If the service wasn't start then he feel renege with probability p, quits the line and not getting the service. If the particular patron appeal for the service with retention tactics he may keep in that same queue by q $(= 1 - p)$. If the system has n patrons then we take the probability as Pn. The time distribution T with parameter $\tilde{\xi}$ accompanies an exponential structure.

By basic queueing theory, let us take

$$\tilde{\lambda}P_0 = \tilde{\mu} \, P_1$$

$$\left[\left(\frac{\tilde{\lambda}}{n+1}\right) + \tilde{\mu} + (n-1)\tilde{\xi} p\right] P_n = [\tilde{\mu} + n\tilde{\xi} p]P_{n+1} + \left(\frac{\tilde{\lambda}}{n}\right) P_{n-1},$$

$$1 \le n \le N-1 \text{ and}$$

$$\left(\frac{\tilde{\lambda}}{N}\right) P_{N-1} = [\tilde{\mu} + (N-1)\tilde{\xi} p]P_N$$

Then

The forecasted patrons in the System $(N_S) = \sum_{n=1}^{N} n P_n$

$$\Rightarrow (N_s) = \sum_{n=1}^{N} n \left[\frac{1}{n!} \prod_{k=1}^{n} \frac{\tilde{\lambda}}{\tilde{\mu} + (k-1)\tilde{\xi} p}\right] P_0 \tag{1}$$

The forecasted patrons Served in the unit time $(E(CS)) = \sum_{n=1}^{N} \tilde{\mu} P_n$

$$\Rightarrow (E(CS)) = \tilde{\mu} \sum_{n=1}^{N} \left[\frac{1}{n!} \prod_{k=1}^{n} \frac{\tilde{\lambda}}{\tilde{\mu} + (k-1)\tilde{\xi} p}\right] P_0 \tag{2}$$

Average Reneging Rate $(Rr) = \sum_{n=1}^{N} (n-1)\tilde{\xi} p P_n$

$$\Rightarrow (Rr) = \sum_{n=1}^{N} (n-1)\tilde{\xi} p \frac{1}{n!} \left[\prod_{k=1}^{n} \frac{\tilde{\lambda}}{\tilde{\mu} + (k-1)\tilde{\xi} p}\right] P_0 \tag{3}$$

Average Retention Rate $(RR) = \sum_{n=1}^{N} (n-1)\tilde{\xi} q P_n$

$$\Rightarrow (RR) = \sum_{n=1}^{N} (n-1)\tilde{\xi} q \frac{1}{n!} \left[\prod_{k=1}^{n} \frac{\tilde{\lambda}}{\tilde{\mu} + (k-1)\tilde{\xi} p}\right] P_0 \tag{4}$$

where $P_0 = \dfrac{1}{1+\sum_{n=1}^{N} \left[\frac{1}{n!} \prod_{k=1}^{n} \frac{\tilde{\lambda}}{\tilde{\mu}+(k-1)\tilde{\xi} p}\right]}$.

(II) *Cost Expedients*

Symbols
R_R = Mean retention rate.
R_r = Mean reneged rate.
C_h = The holding cost/unit time.
C_s = The service cost/unit time.
C_l = The cost for every lost patron/unit time.
C_R = The cost for every retained patron/unit time.
C_r = The cost for every reneged patron/unit time.
R = The cost of revenue by the service for every patron/unit time.
$\lambda_{lost} = \lambda P_N$ = Mean rate for a missed patron.

AEC = Aggregate expected cost.
AEP = Aggregate expected profit.
AER = Aggregate expected revenue.

Now by using (1)–(4)

(i) $\text{AEC} = \tilde{C}_s \tilde{\mu} + \widetilde{C_h} \widetilde{N}_s + \tilde{C}_l \tilde{\lambda} \widetilde{P}_N + \tilde{C}_r \widetilde{R}_r + C_R \widetilde{R}_R$

(ii) $\text{AER} = \tilde{R} \widetilde{N}_s - \tilde{R} \tilde{\lambda} \widetilde{P}_N - \tilde{R} \widetilde{R}_r$

(iii) $\text{AEP} = \text{TER} - \text{T EC}.$

4   Centroid of Centroids Fuzzy Ordering Method—Algorithm

The centroid of the trapezium (APQD) is appraised as the equilibrium prong (Fig. 4).
Split the trapezium A^U as a rectangle (BPQC) and two triangles (APB), (CQD). We
get three centroids $G^U{}_1$, $G^U{}_2$, and $G^U{}_3$ from these plane figures as in Fig. 4. The
centroid $\mathbf{G^U}$ of $G^U{}_1$, $G^U{}_2$, and $G^U{}_3$ is to be the reference point and also the balancing
point which determines the ordering of generalized TrFN. Similarly, in this same
way, we detect G^L of the trapezoid A^L.

Contemplate a generalized trapezoidal fuzzy number $\tilde{A}^U = (aU, bU, cU, dU;h1)$. The
centroids are $G^U{}_1 = \left(\frac{a^U + 2b^U}{3}, \frac{h_1}{3}\right)$; $G^U{}_2 = \left(\frac{b^U + c^U}{2}, \frac{h_1}{2}\right)$ and $G^U{}_3 = \left(\frac{2c^U + d^U}{3}, \frac{h_1}{3}\right)$.
These centroids are non-collinear points also $G^U{}_1$, $G^U{}_2$ and $G^U{}_3$ create a triangle.

Then we catch the centroid
$G^U_{\tilde{A}^U} = \overline{x_0}, \overline{y_0}$ (ie. G^U) of this triangle with the generalized TrFN
$\tilde{A}^U = (aU, bU, cU, dU;h1)$ as

$$G^U_{\tilde{A}^U} = \overline{x_0}, \overline{y_0} = \left(\frac{2a^U + 7b^U + 7c^U + 2d^U}{18}, \frac{7h_1}{18}\right).$$

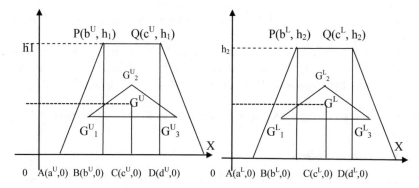

Fig. 4 Centroid of centroids

The ordering function of the generalized TrFN

$$\tilde{A}^U = \left(a^U, b^U, c^U, d^U; h_1\right) \text{ is } R\left(\tilde{A}^U\right) = \bar{x}_0 \times \bar{y}_0$$

$$= \left(\frac{2a^U + 7b^U + 7c^U + 2d^U}{18}\right) X \left(\frac{7h_1}{18}\right)$$

Similarly, the ordering function of the generalized trapezoidal fuzzy number

$$\tilde{A}^L = \left(a^L, b^L, c^L, d^L; h_2\right) \text{ is } R\left(\tilde{A}^L\right) = \bar{x}_0 \times \bar{y}_0$$

$$= \left(\frac{2a^L + 7b^L + 7c^L + 2d^L}{18}\right) X \left(\frac{7h_2}{18}\right)$$

Finally the ordering function of the

$$\text{IVT2FN } \tilde{A} = [A^U, A^L] \text{ is } R(\tilde{A}) = \frac{R\left(\tilde{A}^U\right) + R\left(\tilde{A}^L\right)}{2}.$$

5　Illustration

Now the Government of Tamilnadu announced that, every house holder should link the Aadhar number with their Electricity board number. The government also arranged some special camp for this linking purpose in a free of cost. So every house holder wants to link the Aadhar number and Electricity board number. So in these special camps, a lot of crowds are waiting for to get the service. Here we consider an M/M/1 queue with finite capacity (N). Most of the people are waiting in a long queue and utilized the service. Some of the people may get discourage with these long line.

Let q = 0, 0.1, 0.2, ..., 1, since we take N = 10, the arrival rate $\tilde{\lambda}$ = [(1,3,8,10;1) (2,4,7,9;0.9)], the service rate $\tilde{\mu}$ = [(5,8,12,15;1) (6,9,11,13;0.9)] and time distribution parameter $\tilde{\xi}$ = [(0.1,0.3,0.8,1.0;1) (0.2,0.4,0.7,0.9;0.9)]/hr individually.

By our proposed Fuzzy Ordering approach we got R($\tilde{\lambda}$) = 2.05, R($\tilde{\mu}$) = 3.74, R($\tilde{\xi}$) = 0.2.

(I) *Performance Expedients*

By using the queueing models we caught the performance expedients concerning q.

Table 1 configures that If q = 0, then R_R = 0. It means that no patron has retained. If q = 1 then Rr = 0, it means every reneging patrons have retained. If q gets increments then N_s, **E(CS)** and R_R are getting gradual increments but the R_r gets subsequent decrease.

Figures 5, 6 and 8 are configuring. If q gets increments then N_s, **E(CS)** and R_R are getting gradual increments but the R_r gets subsequent decrease (Fig. 7).

(II) *Cost Expedients*

Let the arrival rate $\tilde{\lambda}$ = [(1,3,8,10;1) (2,4,7,9;0.9)]/hr, the service rate
$\tilde{\mu}$ = [(5,8,12,15;1) (6,9,11,13;0.9)]/hr and time distribution parameter
$\tilde{\xi}$ = [(0.1,0.3,0.8,1.0;1) (0.2,0.4,0.7,0.9;0.9)]/hr.

Also N = 10, $\tilde{C_l}$ = [(100,300,800,1000;1) (200,400,700,900;0.9)], $\tilde{C_s}$ = [(10,30,80,100;1) (20,40,70,90;0.9)], $\tilde{C_h}$ = [(21,23,28,30;1) (22,24,27,29;0.9)], $\tilde{C_r}$ = [(55,75,125,145;1) (65,85,115,135;0.9)] and $\tilde{R}$ = [(550,750,1250,1450;1) (650,850,1150,1350;0.9)].

Table 1 Performance expedients versus q

S. no.	q	N_s	E(CS)	R_r	R_R
i	0	5180×10^{-4}	1642×10^{-3}	207×10^{-4}	0
ii	0.1	5211×10^{-4}	1645×10^{-3}	186×10^{-4}	20×10^{-4}
iii	0.2	5240×10^{-4}	1647×10^{-3}	166×10^{-4}	41×10^{-4}
iv	0.3	5269×10^{-4}	1649×10^{-3}	144×10^{-4}	62×10^{-4}
v	0.4	5299×10^{-4}	1651×10^{-3}	122×10^{-4}	81×10^{-4}
vi	0.5	5330×10^{-4}	1653×10^{-3}	101×10^{-4}	101×10^{-4}
vii	0.6	5361×10^{-4}	1655×10^{-3}	80×10^{-4}	120×10^{-4}
viii	0.7	5392×10^{-4}	1657×10^{-3}	60×10^{-4}	140×10^{-4}
ix	0.8	5421×10^{-4}	1659×10^{-3}	39×10^{-4}	156×10^{-4}
x	0.9	5452×10^{-4}	1662×10^{-3}	19×10^{-4}	171×10^{-4}
xi	1.0	5483×10^{-4}	1664×10^{-3}	0	185×10^{-4}

Fig. 5 q versus Ns

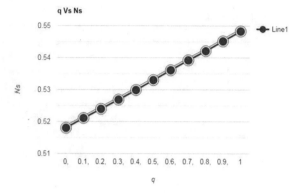

Fig. 6 q versus E(CS)

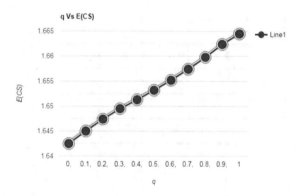

Fig. 7 q versus Rr

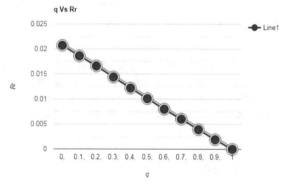

By our proposed Fuzzy Ordering approach we got $R(\widetilde{C}_l) = 205.23$, $R(\widetilde{C}s) = 20.52$, $R(\widetilde{C}_h) = 9.12$, $R(\widetilde{C}_r) = 37.46$, $R(\widetilde{\xi}) = 0.2$, $R(\widetilde{R}) = 374.62$, $R(\widetilde{\lambda}) = 2.05$, $R(\widetilde{\mu}) = 3.74$.

Here we calculated the variations in Cost stratagems w.r.t q.

Fig. 8 q versus R_R

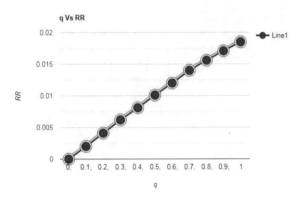

Table 2 Cost expedients versus q

S. no	q	C_R	AEC	AER	AEP
i	0	0	8260×10^{-2}	$18,566 \times 10^{-2}$	$10,360 \times 10^{-2}$
ii	0.1	5	8248×10^{-2}	$18,759 \times 10^{-2}$	$10,511 \times 10^{-2}$
iii	0.2	10	8238×10^{-2}	$18,947 \times 10^{-2}$	$10,709 \times 10^{-2}$
iv	0.3	15	8231×10^{-2}	$19,132 \times 10^{-2}$	$10,901 \times 10^{-2}$
v	0.4	20	8221×10^{-2}	$19,311 \times 10^{-2}$	$11,090 \times 10^{-2}$
vi	0.5	25	8219×10^{-2}	$19,593 \times 10^{-2}$	$11,374 \times 10^{-2}$
vii	0.6	30	8208×10^{-2}	$19,771 \times 10^{-2}$	$11,563 \times 10^{-2}$
viii	0.7	35	8199×10^{-2}	$19,964 \times 10^{-2}$	$11,765 \times 10^{-2}$
ix	0.8	40	8188×10^{-2}	$20,162 \times 10^{-2}$	$11,974 \times 10^{-2}$
x	0.9	45	8177×10^{-2}	$20,353 \times 10^{-2}$	$12,176 \times 10^{-2}$
xi	1.0	50	8165×10^{-2}	$20,540 \times 10^{-2}$	$12,375 \times 10^{-2}$

Table 2 configures that, If q gets increments then AER and AEP are getting gradual increments but the AEC gets subsequent decrease.

Figures 10 and 11 configure that, If q gets increments then AER and AEP are getting gradual increments but the AEC gets subsequent decrease (Fig. 9).

6 Conclusion

We have analyzed with a new Centroid based fuzzy ordering technique in the discouraged arrivals Queueing system in (IVT2FN) attitude. Discouraged arrivals queuing models have been manipulated in many service mechanisms for estimating system attainments in (IVT2FN) manner. The analyzer can clutch the foremost and supreme determinations. This manner brought out in this paper forecasts factual illumination for analyzers. We terminate that the sequel of fuzzy problems can be acquired by

Fig. 9 q versus AEC

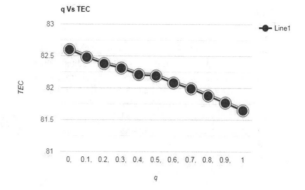

Fig. 10 q versus AER

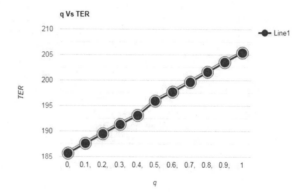

Fig. 11 q versus AEP

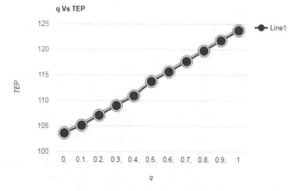

the fuzzy ordering procedure very profitably. Our proposed method is very efficient and more loyal. The verdicts which we got from this method are very strong and concrete. Further, we will endeavor to get the performance expedients for (IVT2FS) by using Incenter, Orthocenter, and Circumcenter-based orderings.

References

1. Allaviranloo, T., Jahantigh, M.A., Hajighasemi, S.: A new distance measure and ranking method for generalized trapezoidal fuzzy numbers. Math. Prob. Eng. **2013**, 6 (2013)
2. Alterio, P.D., Garibaldi, J.M., John, R.I., Pourabdollah, A.: Constrained interval type-2 fuzzy sets. In: IEEE Transactions on Fuzzy Systems, vol. 29, no. 5, pp. 1212–1225, May 2021. https://doi.org/10.1109/TFUZZ.2020.2970911
3. Ammar, S.I., El-Sherbiny, A.A., Al-Seedy, R.O.: A matrix approach for the transient solution of an M/M/1/N queue with discouraged arrivals and reneging. Int. J. Comput. Math. **89**, 482–491 (2012)
4. Azman, F.N., Abdullah, L.: Ranking fuzzy numbers by centroid method. Malaysian J. Fund. Appl. Sci. **8**(3), 117–121 (2012)
5. Bose, S.: An Introduction to Queueing Systems. Kluvar Academic/Plenum Publishers, New Yark (2008)
6. Chu, T.C., Tsao, C.T.: Ranking fuzzy numbers with an area between the centroid point and original point. Comput. Math. Appl. **43**(1–2), 111–117 (2002)
7. Dat, L.Q., Yu, V.F., Chou S.Y.: An improved ranking method for fuzzy numbers based on the centroid index. Int. J. Fuzzy Syst. **14**(3), 413–419 (2012)
8. De, A., Kundu, P., Das, S., et al.: A ranking method based on interval type-2 fuzzy sets for multiple attribute group decision making. Soft Comput. **24**, 131–154 (2020)
9. Figueroa-García, J.C., Chalco-Cano, Y., Román-Flores, H.: Yager index and ranking for interval type-2 fuzzy numbers. In: IEEE Transactions on Fuzzy Systems, vol. 26, no. 5, pp. 2709–2718, October 2018. https://doi.org/10.1109/TFUZZ.2017.2788884
10. Hari Ganesh, A., Helen Shobana, A., Ramesh, R.: Identification of critical path for the analysis of bituminous road transport network using integrated FAHP–FTOPSIS method. Mater. Today: Proc. **37**, 193–206 (2021)
11. Janos, S.: Basic queueing theory. Globe Edit Publishers, Omniscriptum GMBH, Germany (2016)
12. Klir, G.J., Yuvan, B.: Fuzzy Sets and Fuzzy Logic Theory and Applications. Prentice Hall, India (2005)
13. Kumar, R., Sharma, S.K.: M/M/1/N queuing system with retention of reneged customers. Pak. J. Stat. Oper. Res. **8**, 859–866 (2012)
14. Mendel, J.M., John, R.I.B.: Type-2 fuzzy sets made simple. IEEE Trans. Fuzzy Syst. **10**(2), 117–127 (2002)
15. Mendel, J.M., John, R.I., Liu, F.: Interval type-2 fuzzy logic systems made simple. IEEE Trans. Fuzzy Syst. **14**(6), 808–821 (2006)
16. Mitchell, H.B.: Ranking type-2 fuzzy numbers. IEEE Trans. Fuzzy Syst. **14**, 287–294 (2006)
17. Parandin, N., Araghi, M.A.F.: Ranking of fuzzy numbers by distance method. J. Appl. Math., Winter, **15**(19), 47–55
18. Qin, S., Zhang, C., Zhao, T., Tong, W., Bao, Q., Mao, Y.: Dynamic high-type interval type-2 fuzzy logic control for photoelectric tracking system processes, **10**, 562, 1–20. https://doi.org/10.3390/pr10030562
19. Ramesh, R., Hari Ganesh, A.: M/M/1/n fuzzy queueing models with discouraged arrivals under Wingspans fuzzy ranking method. Int. J. Appl. Eng. Res. **14**(4), 1–12 (2019)
20. Rao, P.P.B., Shankar, N.R.: Ranking fuzzy numbers with a distance method using circumcenter of centroids and an index of modality. Adv. Fuzzy Syst. **2011**, ID 178308 (2011)
21. Rao, P.P.B., Shankar, N.R.: Ranking generalized fuzzy numbers using area, mode, spreads and weights. Int. J. Appl. Sci. Eng. **10**(1), 41–57
22. Shankar, N.R., Sarathi, B.P., Babu, S.S.: Fuzzy critical path method based on a new approach of ranking fuzzy numbers using centroid of centroids. Int. J. Fuzzy Syst. Appl. **3**(2), 16–31 (2013)
23. Wang, Y.J., Lee, H.S.: The revised method of ranking fuzzy numbers with an area between the centroid and original points. Comput. Math. Appl. **55**(9), 2033–2042

24. Westman, L., Wang, Z.: Ranking fuzzy numbers by their left and right Wingspans. In: Joint IFSA World Congress and NAFIPS Annual Meeting, pp. 1039–1044. Edmonton (2013)
25. Zhao, M., Qin, S.-S., Li, Q.-W., Lu, F.-Q., Shen, Z.: The likelihood ranking methods for interval type-2 fuzzy sets considering risk preferences. Math. Prob. Eng. **2015**, Article ID 680635, 12 (2015)

Extension of Fuzzy Principal Component Analysis to Type-2 Fuzzy Principal Component Analysis

Daoudi Bouchra, Hamzaoui Hassania, and Mounir Gouiouez

Abstract This chapter compares type 1 and type 2 fuzzy principal component analysis which are based on type 1 and type 2 fuzzy C-means algorithms, respectively. The two clustering methods are the combination of k-means clustering algorithm and type 1 and type 2 fuzzy logic, respectively.

1 Introduction

The availability of data is growing exponentially, which has increased uncertainty rates. On the other hand, real data are not always precise numbers or classes; they are fuzzy data, they consider that all phenomena in the physical universe have an inherent degree of uncertainty; therefore, Zadeh (1921–2017) introduced his fuzzy set theory (1965) as a mathematical way to represent them. He also defined fuzzy logic, which appeared in the context of fuzzy set theory, as a generalization of boolean logic based on understanding the human language system and processing human reasoning. This field has attracted a great deal of interest from the research community in various areas, such as statistics, which is the science of collecting, analyzing, presenting, and interpreting data. Among the most widely used statistical methods is principal component analysis.

The original version of the chapter has been revised: The family name and given name of the third author have been updated. A correction to this chapter can be found at https://doi.org/10.1007/978-3-031-26332-3_17

D. Bouchra (✉) · H. Hassania · M. Gouiouez
LPAIS, Faculty of Sciences, Sidi Mohamed Ben Abdellah University, Fez, Morocco
e-mail: bouchra.daoudi@usmba.ac.ma

H. Hassania
e-mail: hassania.hamzaoui@usmba.ac.ma

The basic idea of the Principal Component Analysis (PCA) is to analyze and explore the set of multidimensional data in order to reduce the dimension of this dataset. This approach is mainly characterized by its performance in the analysis process in which it gathers elements that share the same characteristics. However, it has many problems [1] such as sensitivity to noise, the existence of isolated points and loss of information. To overcome this limitation and make PCA more robust, researchers proposed to use Zadeh's fuzzy set theory [1, 2] as fuzzy principal component analysis (FPCA) [3–12] or type 1 fuzzy principal component analysis (T1FPCA), which uses the fuzzy c-means (FCM) algorithm [13], an unsupervised algorithm generalizing the k-means algorithm. Their procedure is that each point in the data set belongs to each cluster with a certain degree. And also, the fuzzy correlation matrix which is computed using the membership degrees obtained by the FCM algorithm. Others have shown that the type 1 membership function (MF) in FPCA can itself be fuzzy. In this case, we talk about the type 2 fuzzy sets, which are used to handle more uncertainty.

Type 2 fuzzy sets [14–17] are the extension or the generalization of type 1 fuzzy sets [7, 18, 19], they are used in principal component analysis as type 2 fuzzy principal component analysis (T2FPCA) [20, 21]. Which, their membership functions are themselves fuzzy i.e., they are type 1 fuzzy sets. T2FPCA followed the same approach, i.e., they relied on the interval-2 fuzzy c-means (IT2FCM) clustering algorithm [18, 22, 23] and the fuzzy correlation matrix.

Moreover, type 2 fuzzy sets are used in different fields due to their importance, and in these papers [21, 24–32], their high performance is always found.

In this work, we will start with an overview of PCA and its algorithm. And we will present FPCA as a solution to overcome the problems and the limitations of PCA and make it more robust. Then we will put in light the impact that T2FPCA has in this way. Finally, we will compare the results obtained after applying these last two methods to two data sets.

2 PCA

Principal component analysis was first proposed by Hotelling in 1933 as a fundamental method of multidimensional descriptive statistics, it allows to treat quantitative variables. It is known to reduce the dimensionality of the dataset while keeping most of the information. Mathematically, it is a change of basis from the canonical basis representation of the original variables to the component basis representation.

2.1   The Main Steps of the PCA Algorithm

Consider a dataset X of p variables and n observations.

- **Step 1**. Normalize and standardize the dataset: to bring it to a common scale and unit.
  - Normalize the dataset by the following formula:

$$\frac{X - X_{\min}}{X_{\max} - X_{\min}}.$$

  - Standardize the dataset by the following formula:

$$\frac{X - mean}{Std}.$$

- **Step 2**. Compute the inertia matrix by the following formula:

$$C_{p \times p} = \frac{1}{n} \sum_{i=1}^{n} \left(X_i - \overline{X_i}\right)\left(X_i - \overline{X_i}\right)^T.$$

- **Step 3**. Determine the eigenvalues and eigenvectors of the covariance matrix.
- **Step 4**. Projection of dataset on principal components.

3 Type 1 Fuzzy PCA

The T1FPCA is based on the FCM clustering algorithm to find membership degrees.

The main objective of FCM [13] is to obtain the fuzzy partition for the dataset $X = (x_1, x_2, \ldots, x_n)$ into p fuzzy clusters, and the cluster centers by minimizing the function J_m.

$$J_m(U, C) = \sum_{i=1}^{n} \sum_{j=1}^{p} \left(\mu_{ij}\right)^m \left\| x_i - c_j \right\|^2$$

where $- U = \left(\mu_{ij}\right)_{n \times p}, (p \le n)$, is a fuzzy partition matrix, $\mu_{ij} \in [0, 1]$ is the membership degree of data point x_i to the fuzzy cluster j.

  $- C = (c_1, c_2, \ldots, c_p)$ are the cluster centers. c_j is the jth cluster center of the fuzzy cluster.

  $- \left\| x_i - c_j \right\|$ is the Euclidean norm (distance) between x_i and c_j.

  $- m > 1$ fuzziness parameter or the fuzzifier, it controls the fuzzy degree of membership of each data (The particular case that $m = 1$ corresponds to the k-means algorithm). The larger m is, the more fuzzy the partition is.

3.1 The Steps for the T1FPCA

T1FPCA Summary: The first phase is to obtain the membership matrix from the FCM clustering algorithm and the second phase is to calculate the fuzzy correlation matrix and determine its eigenvalues and eigenvectors.

Firstly, we fix p the number of clusters, the tolerance value ε, the maximum number of iterations $Iter\_max$, and choose the fuzziness parameter m.

- **Step 1**. Initialization of the elements of the membership matrix.
  $\forall i = 1, \ldots, n. \ \forall j = 1, \ldots, p. \ \mu_{ij} \in [0, 1]$ such that:

$$\forall i = 1, \ldots, n \qquad \sum_{j=1}^{c} u_{ij} = 1, \tag{1}$$

- **Step 2**. determination of cluster centers c_j by the formula:
  for $j = 1, 2, \ldots, p$.

$$c_j = \frac{\sum_{i=1}^{n} \mu_{ij}^m \cdot x_i}{\sum_{i=1}^{n} \mu_{ij}^m}, \tag{2}$$

- **Step 3**. Membership matrix updating, such they satisfy the constraint (1) by the formula:

$$\mu_{ij} = \left(\sum_{k=1}^{p} \left(\frac{\|x_i - c_j\|}{\|x_i - c_k\|} \right)^{2/(m-1)} \right)^{-1} , i = 1, \ldots, n \ \text{ and } \ j = 1, \ldots, p. \tag{3}$$

- **Step 4**. If $\left\| U^{(k+1)} - U^{(k)} \right\| \leq \varepsilon$, where k is the iteration step. Or $Iter\_max$ reached we go to step 5 otherwise we repeat steps 2 and 3.
- **Step 5**. Calculate the covariance matrix M using the fuzzy membership matrix determined above.

$$M_{kl} = \frac{\sum_{i=1}^{n} u_{ij}^m (x_{ik} - \bar{x}_k)(x_{il} - \bar{x}_l)}{\sum_{i=1}^{n} u_{ij}^m} \tag{4}$$

- **Step 6**. Determine the eigenvalues and eigenvectors of the covariance matrix M as usual;

4 Type 2 Fuzzy PCA

In T2FPCA [2], we use the IT2FCM clustering algorithm [7, 24, 33], to obtain the MF and the clustering centers (The same as the T1FPCA). We will apply the following procedure.

To obtain the fuzzy partition for the dataset $X = (x_1, x_2, \ldots, x_n)$ into p fuzzy clusters and all the parameters to be extracted from the IT2FCM clustering algorithm, two different objective functions are considered J_{m_1} and J_{m_2} to minimize.

$$J_{m_1}(U, C) = \sum_{i=1}^{n} \sum_{j=1}^{P} \mu_j (x_i)^{m_1} d_{ij}^2 \tag{5}$$

$$J_{m_2}(U, C) = \sum_{i=1}^{n} \sum_{j=1}^{P} u_j (x_i)^{m_2} d_{ij}^2 \tag{6}$$

where $- C = (c_1, c_2, \ldots, c_p)$ are the cluster centers.

$- U = (\mu_{ij})_{n \times p}$, $(p \leq n)$, is a fuzzy partition matrix, $\mu_{ij} \in [0, 1]$ is the membership degree of data point x_i to the fuzzy cluster j.

$- m_1$ and m_2 are two different fuzziness parameters (i.e. fuzzifiers). $m_1, m_2 \geq 1$.

$- d_{ij}$: the Euclidean distance between a point x_i and the prototype of the jth cluster.

Using the Lagrange multiplier function to rewrite the objective functions given in (5) and (6) we find:

$$J (U, C, m_1, m_2, \alpha, \gamma) = \sum_{i=1}^{n} \sum_{j=1}^{p} \mu_j (x_i)^{m_1} d_{ij}^2 + \gamma \sum_{i=1}^{n} \sum_{j=1}^{p} \mu_j (x_i)^{m_2} d_{ij}^2$$
$$+ \sum_{i=1}^{n} \alpha_i \left[\sum_{j=1}^{p} \mu_j (x_i) - 1 \right] \tag{7}$$

where the coefficients α and γ are the parameters of the Lagrange multiplier.

4.1 The Steps for the T2FPCA

T2FPCA Summary: the IT2FCM algorithm is applied to the dataset X to obtain the membership matrix and use it to compute the fuzzy mean and the fuzzy correlation matrix and determine its values and eigenvectors.

Firstly, we fix p the number of clusters, the tolerance value ε, the maximum number of iterations $Iter\_max$, and choose the fuzziness parameters m_1 and m_2:

- **Step 1**. Initialize the centroids.
- **Step 2**. Calculate the upper and the lower memberships matrix by the formulas:
  for $i = 1, \ldots, n$ and $j = 1, \ldots, p$

$$\bar{\mu}_{ij}(\mathbf{x}_i) = \begin{cases} \dfrac{1}{\sum\limits_{k=1}^{p} \left(\dfrac{d_{ij}}{d_{ki}}\right)^{\frac{2}{(m_1-1)}}}, & \text{if } \dfrac{1}{\sum\limits_{k=1}^{p} \left(\dfrac{d_{ij}}{d_{ik}}\right)} < \dfrac{1}{p} \\[4mm] \dfrac{1}{\sum\limits_{k=1}^{p} \left(\dfrac{d_{ij}}{d_{ki}}\right)^{\frac{2}{(m_2-1)}}}, & \text{otherwise} \end{cases} . \tag{8}$$

$$\underline{\mu}_{ij}(\mathbf{x}_i) = \begin{cases} \dfrac{1}{\sum\limits_{k=1}^{p} \left(\dfrac{d_{ij}}{d_{ki}}\right)^{\frac{2}{(m_1-1)}}}, & \text{if } \dfrac{1}{\sum\limits_{k=1}^{p} \left(\dfrac{d_{ij}}{d_{ik}}\right)} \geq \dfrac{1}{p} \\[4mm] \dfrac{1}{\sum\limits_{k=1}^{p} \left(\dfrac{d_{ij}}{d_{ki}}\right)^{\frac{2}{(m_2-1)}}}, & \text{otherwise} \end{cases} . \tag{9}$$

such they satisfy the constraint (1).

- **Step 3**. Calculate the cluster centers:

$$c_i = \frac{c_i^L + c_i^R}{2}, \quad i = 1, \ldots, c.$$

Such as the left value c^L and a right value c^R are obtained by the Karnik and Mendel (KM) iterative algorithm (in the Sect. 4.1.1) [14, 34, 35].

- **Step 4**. If $\left\| U^{(k+1)} - U^{(k)} \right\| \leq \varepsilon$, where k is the iteration step and

$$\mu_i(x_i) = \frac{\bar{\mu}_{ij}(x_i) + \underline{\mu}_{ij}(x_i)}{2}, \quad i = 1, \ldots, n \text{ and } j = 1, \ldots, p.$$

Or $Iter\_max$ reached we go to step 5 otherwise we repeat steps 2 and 3.

- **Step 5**. Compute the fuzzy mean of the data (u_P):

$$u_P = \frac{\sum_{i=1}^{n} \mu_i x_i}{\sum_{i=1}^{n} \mu_i} \tag{10}$$

- **Step 6**. Compute the fuzzy covariance matrix (C_P)

$$C_P = \frac{1}{n} \sum_{i=1}^{n} \mu_i^2 (x_i - u_P)(x_i - u_P)^T \tag{11}$$

- **Step 7**. Determine the eigenvalues and eigenvectors of the fuzzy covariance matrix C_p as usual.

4.1.1   The KM Iterative Algorithm for Finding the Centers

- **Step 1**. Choosing the fuzzifier m arbitrary. For example $m = 2$.
  Compute centroid $c_j^{(t)} = (c_{j1}, \ldots, c_{jM})$, $j = 1, \ldots, p$. by (2) such as

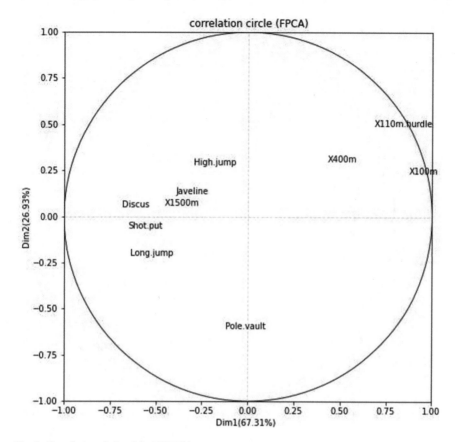

Fig. 1 Correlation circle of the T1FPCA

$$\mu_{ij}(t) = \frac{\overline{\mu}_j(x_i) + \underline{\mu}_j(x_i)}{2}, \quad i = 1, \ldots, n \text{ and } j = 1, \ldots, p.$$

Table 1 The eigenvalues and their variability (the first five)

	T1FPCA		T2FPCA	
Component	Eigenvalue	Variability(%)	Eigenvalue	Variability(%)
1	6.73	67.31	10	99.99
2	2.38	21.65	1.77×10^{-15}	1.77×10^{-14}
3	0.43	4.37	4.77×10^{-16}	4.77×10^{-15}
4	0.13	1.34	1.25×10^{-16}	1.25×10^{-15}
5	0.002	0.02	2.87×10^{-17}	2.87×10^{-16}

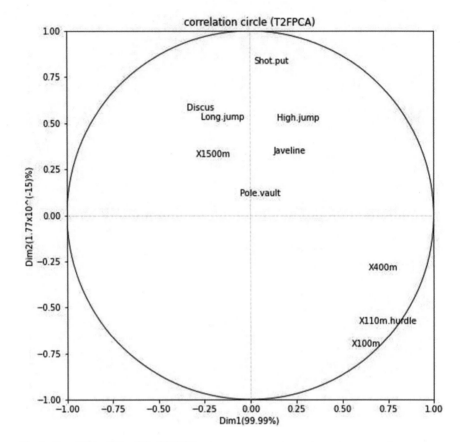

Fig. 2 Correlation circle of the T2FPCA

- **Step 2.** Sort pattern indexes for all n patterns ($i = 1, \ldots, n$) in each of M features ($l = 1, \ldots, M$) in ascending order.
  (i.e., Sorted feature $1 : x_{11} \leq \cdots \leq x_{n1}$

  $$\vdots$$

  Sorted feature $M : x_{1M} \leq \cdots \leq x_{nM}$)
- **Step 3.** Find interval index k ($1 \leq k \leq n - 1$) such that $x_{kl} = c_{jl} = x_{(k+1)l}$.

- **Step 4.** Update $\mu_{ij}^{(t+1)}$ and $c_j^{(t+1)}$:

$$\mu_{ij}^{(t+1)} = \begin{cases} \underline{\mu}_{ij} & k \leq i \\ \bar{\mu}_{ij} & k > i \end{cases} \quad i = 1, \ldots, n \text{ and } j = 1, \ldots, p. \tag{12}$$

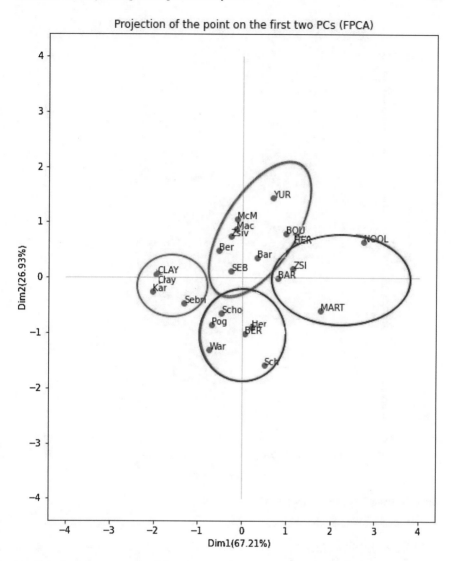

Fig. 3 Individuals scatter plot for T1FPCA

$$c_j^{(t+1)} = \frac{\sum_{i=1}^{n} \left(\mu_{ij}^{(t+1)}\right)^m \cdot x_i}{\sum_{i=1}^{n} \left(\mu_{ij}^{(t+1)}\right)^m}, \quad j = 1, \ldots, p. \tag{13}$$

- **Step 5**. If $c_j^{(t+1)}$ is equal to $c_j^{(t)}$, the procedure is terminated. Otherwise move to step 3

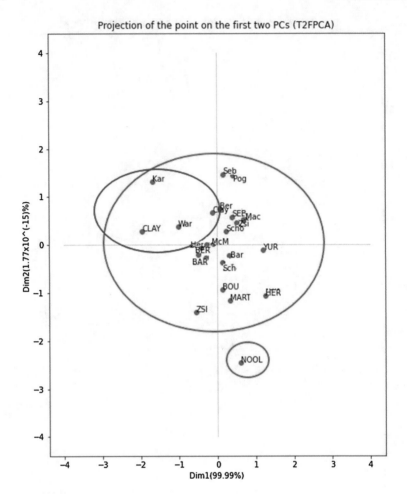

Fig. 4 Individuals scatter plot for T2FPCA

Get $c_j^R = c_j^{(t+1)} = (c_{j1}, \ldots, c_{jM})$, $j = 1, \ldots, p$.

In the case of calculating c_j^L, the procedure is the same as above, but (14) has to be used instead of (12) in step 4,

$$\mu_{ij}^{(t+1)} = \begin{cases} \bar{\mu}_{ij} & k \leq i \\ \underline{\mu}_{ij} & k > i \end{cases} \quad i = 1, \ldots, n \text{ and } j = 1, \ldots, p. \tag{14}$$

Table 2 The eigenvalues and their variability (the first five)

Component	T1FPCA		T2FPCA	
	Eigenvalue	Variability(%)	Eigenvalue	Variability(%)
1	8.58	78.02	10.99	99.99
2	2.69	26.93	1.77×10^{-15}	1.61×10^{-14}
3	0.02	0.26	3.73×10^{-16}	3.39×10^{-15}
4	0.006	0.05	2.13×10^{-16}	1.94×10^{-15}
5	0.0003	0.002	1.36×10^{-16}	1.24×10^{-15}

5 Numerical Comparison of T1FPCA to T2FPCA

In this section, we will numerically compare the two data analysis techniques, type 1 fuzzy PCA and type 2 fuzzy PCA by applying them to two real examples.

Example 1 In this example, we use the dataset decathlon2, which describes the performances of the athletes during two sports events (Desctar and OlympicG), it contains 27 individuals (athletes) described by 13 variables, but in our case, we will only use the individuals and the active variables (the first 10 variables and the first 24 individuals).

The results are presented in the following table.

For T1FPCA (Table 1), the first two PCs account for 67.31% and 26.93%, respectively, of the total variation in the dataset. A two-component model, therefore, explains 94.24% of the total variance.

However, in the case of T2FPCA, the results obtained are very different from the results of T1FPCA. For example, the first principal component explains 99.99% of the total variance, i.e. only one component is sufficient but we add the second component for better visualization. In other words, the two-dimensional scatterplot of the 23 individuals given in Fig. 4 is a very good approximation of the original scatterplot in a ten dimensional space. This result shows the ability of T2FPCA to obtain a better fit of the first principal direction among all data. Therefore, the components derived from T2FPCA explain a much larger share of the variance than their T1FPCA counterparts.

Even if, we consider the first factorial plan for T2FPCA and T1FPCA (Figs. 1, 2, 3 and 4).

The correlation circles (Figs. 1 and 2) show the relationships between all the variables (positively and negatively correlated variables) and the quality of the variable representation.

For T2FPCA, we notice that there is a strong correlation between the variables and for the quality of the representation of the variables we have: $\times 100$ m, $\times 110$ m.hurdle, $\times 400$ m for example are well represented.

For the T1FPCA, we notice that the variables are scattered around the two factorial axes, but they are more concentrated around the second axis and there is a correlation between the variables ($\times 100$, long jump and $\times 110$.hurdle) and between (javelin, high jump and $\times 1500$ m).

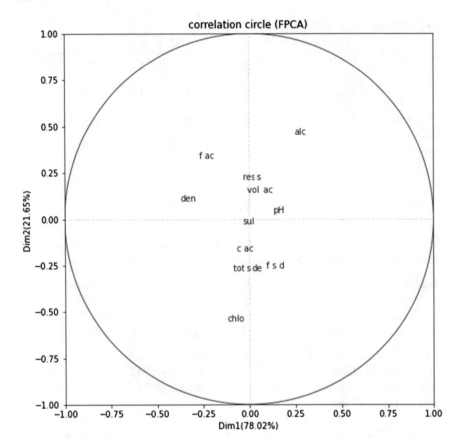

Fig. 5 Correlation circle of the T1FPCA

On the score graph (Figs. 3 and 4) described by PC1 and PC2, i.e., on the first factorial plane (in the case of T1FPCA and T2FPCA) the individuals are well classified in both cases, they define the homogeneous classes (four for T1FPCA and two to three for T2FPCA).

The individuals, in the case of T2FPCA (Fig. 4), are very close to each other, they are dispersed around the first factorial axis, this is logical since the first axis presents 99.99% of the total variance. i.e., we can represent the scatterplot on the graph of the distribution of the individuals by the first factorial axis only. On the contrary, in the case of the T1FPCA (Fig. 4), the individuals are dispersed around the two factorial axes. And this is also normal because the first PC expresses only 67.31% of the total variance.

If we combine successively the correlation circle (Figs. 1 and 2) with the observation graph (Figs. 3 and 4), we can extract the same information (in the case of T2FPCA, we only need the first PC to represent the variables and the individuals, but in the case of type 1 FPCA we need two axes).

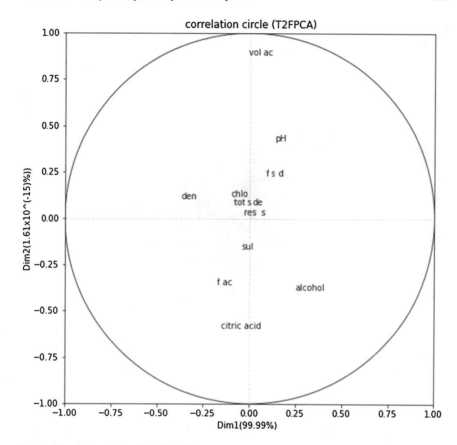

Fig. 6 Correlation circle of the T2FPCA

Example 2: wine quality-red dataset In this example, we use the dataset wine quality-red: which is the quality of red wine based on physicochemical tests. It contains 1599 individuals described by 12 variables, but in our case, we will only use the first 11 variables.

The results are presented in the following table.

We see that the difference between the results obtained by applying the T2FPCA and the T1FPCA (Table 2) is the same as in the previous example.

For T2FPCA, the 1st PC explains 99.99% of the variance in total, compared to 78.02% for T1FPCA, i.e., T2FPCA approximates the original scatterplot in a 10-dimensional space very well only by the 1st PC.

Even so, we consider the first factorial design for T2FPCA and T1FPCA (Figs. 5 and 6).

The correlation circles (Figs. 5 and 6) also show in this example that the variables are well represented by the first factorial axis of T2FPCA. And also we notice that in the case of the T1FPCA, the variables are dispersion on both axes, but they are more

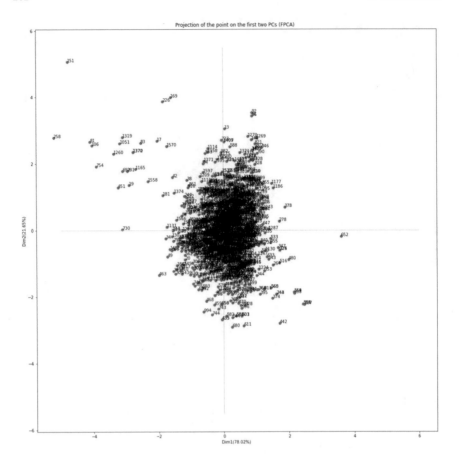

Fig. 7 Individuals scatter plot for T1FPCA

concentrated on the first principal component although the variance expressed by the first axis is not large enough (only 78.02%).

It is difficult to interpret the graph in the case of large samples (Fig. 8), but the dispersion of the individuals around the first principal component of T2PCA is clear. They are very close to each other, i.e., they are very well classified (it represents one class).

And in the case of T1FPCA, the scatter plot of individuals (Fig. 7) described by PC1 and PC2 illustrates a dispersion on both axes, but they are more concentrated on the first principal component.

If we combine successively the correlation circle (Figs. 5 and 6) with the observation graph (Figs. 7 and 8), we can extract the same information (in the case of type 2 FPCA, we only need the first PC to represent the variables and the individuals and in T1FPCA, they are more concentrated on the first principal component).

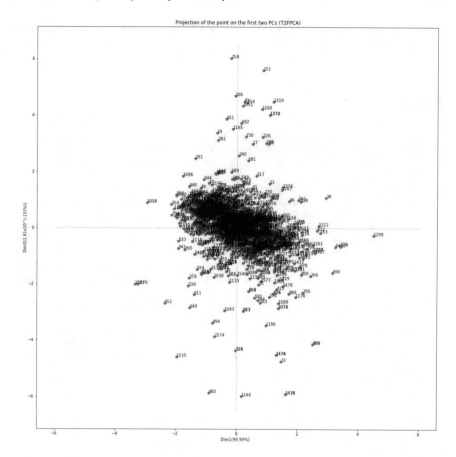

Fig. 8 Individuals scatter plot for T2FPCA

6 Conclusion

We know that PCA has been able to extract very interesting information from a multi-dimensional dataset, using simple graphs, but the major problem with this algorithm is the isolation of points and the existence of noise. This is why the researchers used fuzzy set theory to deal with these problems and to obtain a robust estimation of the PCA.

In this chapter, we presented a detailed study of T1FPCA and T2FPCA and a comparison between these two algorithms. The efficiency of the type 2 fuzzy PCA algorithm was illustrated on two datasets (decatlon2 and wine quality-red). For the T1FPCA method, we seek a fuzzy partition of the dataset into fuzzy clusters. T1FPCA uses the FCM clustering algorithm to determine the membership function. The T2FPCA method uses the IT2FCM algorithm to construct the cluster centers of the fuzzy partition of the dataset and the membership function for these clusters. We applied both methods to real data. We constated that T2FPCA improves T1FPCA and T2FPCA retains more information from X than T1FPCA.

References

1. Jolliffe, I.T.: Principal Component Analysis, 2nd Edn. Springer Series in Statistics (2002)
2. Zadeh, L.A.: Fuzzy Sets (1965) Information and Control, vol. 8, pp. 338–353 (1965)
3. Heoa, G., Gadera, P., Frigui, H.: RKF-PCA: robust kernel fuzzy PCA. Neural Netw. **22**, 642–650 (2009). (Elsevier)
4. Elbanby, Gh., El Madbouly, E., Abdalla, A.: Fuzzy principal component analysis for sensor fusion. In: The 11th International Conference on Information Sciences, Signal Processing and their Applications: Main Tracks
5. Gueorguieva, N., Valova, I., Georgiev, G.: Fuzzyfication of principle component analysis for data dimensionalty reduction. In: IEEE International Conference on Fuzzy Systems FUZZ-IEEE 2016. Published online 2016, pp. 1818–1825 (2016)
6. Pop, H.F., Einax, J.W., Sârbu, C.: Classical and fuzzy principal component analysis of some environmental samples concerning the pollution with heavy metals. Chemom. Intell. Lab. Syst. **97**, 25–32 (2009). (Elsevier)
7. Zimmermann, H.-J.: Fuzzy set theory. In: Advanced Review, WIREs Computational Statistics, vol. 2, no. 3, pp. 317–332 (2010). (May/June 2010)
8. Khanmirza, E., Nazarahari, M., Mousavi, A.: Identification of piecewise affine systems based on fuzzy PCA-guided robust clustering technique. EURASIP J. Adv. Signal Process (2016)
9. Salgado, P., Gonçalves, L., Igrejas, G.: Sliding PCA fuzzy clustering algorithm. In: AIP Conference Proceedings, vol. 1389, pp. 1992 (2011)
10. Hadri, A., Chougdali, K., Touahni, R.: Intrusion detection system using PCA and Fuzzy PCA techniques. In: International Conference on Advanced Communication Systems and Information Security (ACOSIS) (2016)
11. Cundari, T.R., Sarbu, C., Pop, H.F.: Robust fuzzy principal component analysis (FPCA). A comparative study concerning interaction of carbon-hydrogen bonds with molybdenum-oxo bonds. J. Chem. Inf. Comput. Sci. **42**(6), 1363–1369 (2002). (Nov 2002)
12. Xiaohong, W., Jianjiang, Z.: Fuzzy principal component analysis and its kernel-based model. J. Electron. (China) **24**, 772–775 (2007)
13. Nascimento, S., Mirkin, B., Moura-Pires, F.: A fuzzy clustering model of data and fuzzy c-means. In: Ninth IEEE International Conference on Fuzzy Systems. FUZZ-IEEE 2000 (2000)
14. Zhai, D., Mendel, J.M.: Uncertainty measures for general Type-2 fuzzy sets. Inf. Sci. **181**, 503–518 (2011). (Elsevier)
15. Mendel, J.M.: Type 2 fuzzy sets and systems: an overview. IEEE Comput. Intell. Mag. (2007). (Feb 2007)
16. Mendel, J., John, R.: Type-2 fuzzy sets made simple. IEEE Trans. Fuzzy Syst. **10**(2) (2002). (Apr 2002)
17. Nie, M., Tan, W.W.: Modeling capability of type-1 fuzzy set and interval type-2 fuzzy set. IEEE World Congress Comput. Intell. (2012). (10–15 Jun 2012)
18. Kim, E., Oh, S., Pedrycz, W.: Design of reinforced interval type-2 fuzzy c-means-based fuzzy classifier. IEEE Trans. Fuzzy Syst. 1063–6706 (c) (2017)
19. Fathy, E.: A new method for solving the linear programming problem in an interval-valued intuitionistic fuzzy environment. Alexandria Eng. J. **61**(12), 10419–10432 (2022)
20. Singh, V., Verma, N.K., Cui, Y.: Type-2 fuzzy PCA Approach in extracting salient features for molecular cancer diagnostics and prognostics. IEEE Trans. Nanobioscience **18**(3), 482–489 (2019)
21. Taghikhani, S., Baroughi, F., Alizadeh, B.: A generalized interval type-2 fuzzy random variable based algorithm under mean chance value at risk criterion for inverse 1-median location problems on tree networks with uncertain costs. J. Comput. Appl. Math. **408**, 114104 (2022)

22. Hwang, Ch., Chung-Hoon Rhee, F.: Uncertain fuzzy clustering: interval type-2 fuzzy approach to C-means. IEEE Trans. Fuzzy Syst. **15**(1) (2007). (Feb 2007)
23. Linda, O., Manic, M.: General type-2 fuzzy c-means algorithm for uncertain fuzzy clustering. IEEE Trans. Fuzzy Syst. **20**(5) (2012). (Oct 2012)
24. Aminifar, S.: Uncertainty avoider interval type II defuzzification method. Math. Probl. Eng. (2020)
25. Ding, W., Abdel-Basset, M., Hawash, H., Mostafa, N.: Interval type-2 fuzzy temporal convolutional autoencoder for gait-based human identification and authentication. Inf. Sci. (Ny) **597**, 144–165 (2022)
26. Gölcük, I.: An interval type-2 fuzzy axiomatic design method: a case study for evaluating blockchain deployment projects in supply chain. Inf. Sci. (Ny) **602**, 159–183 (2022)
27. Hefaidh, H., Mébarek, D.: Using fuzzy-improved principal component analysis (PCA-IF) for ranking of major accident scenarios. Arab. J. Sci. Eng. **45**(3), 2235–2245 (2020)
28. Rajati, M.R., Mendel, J.M.: Uncertain knowledge representation and reasoning with linguistic belief structures. Inf Sci (Ny) **585**, 471–497 (2022)
29. Singh, V., Verma, N.K., Cu, Y.: Type-2 Fuzzy PCA approach in extracting salient features for molecular cancer diagnostics and prognostics. IEEE Trans. Nanobioscience **18**(3) (2019). (July 2019)
30. Wang, Y., Chen, L., Zhou, J., Li, T., Chen, C.L.P.: Interval type-2 outlier-robust picture fuzzy clustering and its application in medical image segmentation. Appl. Soft. Comput. **122**, 108891 (2022)
31. Wu, L., Qian, F., Wang, L., Ma, X.: An improved type-reduction algorithm for general type-2 fuzzy sets. Inf. Sci. (Ny). **593**, 99–120 (2022)
32. Yan, S.R., Alattas, K.A., Bakouri, M., et al.: Generalized type-2 fuzzy control for type-I diabetes: analytical robust system. Mathematics **10**(5), 1–20 (2022)
33. Chiao, K.P.: The general type 1 and interval type 2 fuzzy sets addition based on the Yager T-norms with entropy as degree of fuzziness. In: 2019 International Conference Fuzzy Theory its Application iFUZZY 2019. Published online, vol. 2019, pp. 214–219 (2019)
34. Mendel, J.M.: On KM algorithms for solving type-2 fuzzy set problems. IEEE Trans. Fuzzy Syst. **21**(3) (2013). (June 2013)
35. Karnik, N.N., Mendel, J.M.: Centroid of a type-2 fuzzy set. Inf. Sci. **132**(2001), 195–220 (2001). (Elsevier)

Correction to: Extension of Fuzzy Principal Component Analysis to Type-2 Fuzzy Principal Component Analysis

Daoudi Bouchra, Hamzaoui Hassania, and Mounir Gouiouez

Correction to:
Chapter 16 in: O. Castillo and A. Kumar (eds.), *Recent Trends on Type-2 Fuzzy Logic Systems: Theory, Methodology and Applications*, **Studies in Fuzziness and Soft Computing 425, https://doi.org/10.1007/978-3-031-26332-3_16**

The original version of this chapter was inadvertently published with incorrect first and last names of the third author which have now been interchanged. The correct name is Mounir Gouiouez. The book and the chapter have been updated with the changes.

The updated version of this chapter can be found at
https://doi.org/10.1007/978-3-031-26332-3_16

© The Author(s), under exclusive license to Springer Nature Switzerland AG 2024
O. Castillo and A. Kumar (eds.), *Recent Trends on Type-2 Fuzzy Logic Systems: Theory, Methodology and Applications*, Studies in Fuzziness and Soft Computing 425,
https://doi.org/10.1007/978-3-031-26332-3_17

Printed in the United States
by Baker & Taylor Publisher Services